INTRODUCTION TO COMMUTATIVE ALGEBRA

This book is in the

ADDISON–WESLEY SERIES IN MATHEMATICS

Consulting Editor: LYNN H. LOOMIS

Introduction to Commutative Algebra

M. F. ATIYAH, FRS
I. G. MACDONALD
UNIVERSITY OF OXFORD

The Advanced Book Program

CRC Press
Taylor & Francis Group
Boca Raton London New York

CRC Press is an imprint of the
Taylor & Francis Group, an **informa** business
A CHAPMAN & HALL BOOK

First published 1969 by Westview Press

Published 2018 by CRC Press
Taylor & Francis Group
6000 Broken Sound Parkway NW, Suite 300
Boca Raton, FL 33487-2742

CRC Press is an imprint of the Taylor & Francis Group, an informa business

Visit the Taylor & Francis Web site at
http://www.taylorandfrancis.com

and the CRC Press Web site at
http://www.crcpress.com

ISBN 13: 978-0-201-40751-8 (pbk)
ISBN 13: 978-0-201-00361-1 (hbk)

Contents

Introduction

Commutative algebra is essentially the study of commutative rings. Roughly speaking, it has developed from two sources: (1) algebraic geometry and (2) algebraic number theory. In (1) the prototype of the rings studied is the ring $k[x_1, \ldots, x_n]$ of polynomials in several variables over a field k; in (2) it is the ring $\mathbf{Z}$ of rational integers. Of these two the algebro-geometric case is the more far-reaching and, in its modern development by Grothendieck, it embraces much of algebraic number theory. Commutative algebra is now one of the foundation stones of this new algebraic geometry. It provides the complete local tools for the subject in much the same way as differential analysis provides the tools for differential geometry.

This book grew out of a course of lectures given to third year undergraduates at Oxford University and it has the modest aim of providing a rapid introduction to the subject. It is designed to be read by students who have had a first elementary course in general algebra. On the other hand, it is not intended as a substitute for the more voluminous tracts on commutative algebra such as Zariski–Samuel [4] or Bourbaki [1]. We have concentrated on certain central topics, and large areas, such as field theory, are not touched. In content we cover rather more ground than Northcott [3] and our treatment is substantially different in that, following the modern trend, we put more emphasis on modules and localization.

The central notion in commutative algebra is that of a prime ideal. This provides a common generalization of the primes of arithmetic and the points of geometry. The geometric notion of concentrating attention "near a point" has as its algebraic analogue the important process of *localizing* a ring at a prime ideal. It is not surprising, therefore, that results about localization can usefully be thought of in geometric terms. This is done methodically in Grothendieck's theory of *schemes* and, partly as an introduction to Grothendieck's work [2], and partly because of the geometric insight it provides, we have added schematic versions of many results in the form of exercises and remarks.

The lecture-note origin of this book accounts for the rather terse style, with little general padding, and for the condensed account of many proofs. We have resisted the temptation to expand it in the hope that the brevity of our presentation will make clearer the mathematical structure of what is by now an elegant and attractive theory. Our philosophy has been to build up to the main theorems in a succession of simple steps and to omit routine verifications.

Anyone writing now on commutative algebra faces a dilemma in connection with homological algebra, which plays such an important part in modern developments. A proper treatment of homological algebra is impossible within the confines of a small book: on the other hand, it is hardly sensible to ignore it completely. The compromise we have adopted is to use elementary homological methods—exact sequences, diagrams, etc.—but to stop short of any results requiring a deep study of homology. In this way we hope to prepare the ground for a systematic course on homological algebra which the reader should undertake if he wishes to pursue algebraic geometry in any depth.

We have provided a substantial number of exercises at the end of each chapter. Some of them are easy and some of them are hard. Usually we have provided hints, and sometimes complete solutions, to the hard ones. We are indebted to Mr. R. Y. Sharp, who worked through them all and saved us from error more than once.

We have made no attempt to describe the contributions of the many mathematicians who have helped to develop the theory as expounded in this book. We would, however, like to put on record our indebtedness to J.-P. Serre and J. Tate from whom we learnt the subject, and whose influence was the determining factor in our choice of material and mode of presentation.

REFERENCES

1. N. Bourbaki, *Algèbre Commutative*, Hermann, Paris (1961–65).
2. A. Grothendieck and J. Dieudonné, Éléments de Géometrie Algébrique, *Publications Mathématiques de l'I.H.E.S.*, Nos. 4, 8, 11, .., Paris (1960–).
3. D. G. Northcott, *Ideal Theory*, Cambridge University Press (1953).
4. O. Zariski and P. Samuel, *Commutative Algebra* I, II, Van Nostrand, Princeton (1958, 1960).

Notation and Terminology

Rings and modules are denoted by capital italic letters, elements of them by small italic letters. A field is often denoted by k. Ideals are denoted by small German characters. **Z**, **Q**, **R**, **C** denote respectively the ring of rational integers, the field of rational numbers, the field of real numbers and the field of complex numbers.

Mappings are consistently written on the *left*, thus the image of an element x under a mapping f is written $f(x)$ and not $(x)f$. The composition of mappings $f: X \to Y$, $g: Y \to Z$ is therefore $g \circ f$, not $f \circ g$.

A mapping $f: X \to Y$ is *injective* if $f(x_1) = f(x_2)$ implies $x_1 = x_2$; *surjective* if $f(X) = Y$; *bijective* if both injective and surjective.

The end of a proof (or absence of proof) is marked thus ■.

Inclusion of sets is denoted by the sign $\subseteq$. We reserve the sign $\subset$ for *strict* inclusion. Thus $A \subset B$ means that A is contained in B and is not equal to B.

1

Rings and Ideals

We shall begin by reviewing rapidly the definition and elementary properties of rings. This will indicate how much we are going to assume of the reader and it will also serve to fix notation and conventions. After this review we pass on to a discussion of prime and maximal ideals. The remainder of the chapter is devoted to explaining the various elementary operations which can be performed on ideals. The Grothendieck language of schemes is dealt with in the exercises at the end.

RINGS AND RING HOMOMORPHISMS

A *ring* A is a set with two binary operations (addition and multiplication) such that

1) A is an abelian group with respect to addition (so that A has a zero element, denoted by 0, and every $x \in A$ has an (additive) inverse, $-x$).
2) Multiplication is associative ($(xy)z = x(yz)$) and distributive over addition ($x(y + z) = xy + xz$, $(y + z)x = yx + zx$).

We shall consider only rings which are *commutative*:

3) $xy = yx$ for all $x, y \in A$,

and have an *identity element* (denoted by 1):

4) $\exists 1 \in A$ such that $x1 = 1x = x$ for all $x \in A$.

The identity element is then unique.

Throughout this book the word "ring" shall mean a commutative ring with an identity element, that is, a ring satisfying axioms (1) to (4) above.

Remark. We do not exclude the possibility in (4) that 1 might be equal to 0. If so, then for any $x \in A$ we have

$$x = x1 = x0 = 0$$

and so A has only one element, 0. In this case A is the *zero ring*, denoted by 0 (by abuse of notation).

A *ring homomorphism* is a mapping f of a ring A into a ring B such that

i) $f(x + y) = f(x) + f(y)$ (so that f is a homomorphism of abelian groups, and therefore also $f(x - y) = f(x) - f(y)$, $f(-x) = -f(x)$, $f(0) = 0$),

ii) $f(xy) = f(x)f(y)$,

iii) $f(1) = 1$.

In other words, f respects addition, multiplication and the identity element.

A subset S of a ring A is a *subring* of A if S is closed under addition and multiplication and contains the identity element of A. The identity mapping of S into A is then a ring homomorphism.

If $f: A \to B$, $g: B \to C$ are ring homomorphisms then so is their composition $g \circ f: A \to C$.

IDEALS. QUOTIENT RINGS

An *ideal* $\mathfrak{a}$ of a ring A is a subset of A which is an additive subgroup and is such that $A\mathfrak{a} \subseteq \mathfrak{a}$ (i.e., $x \in A$ and $y \in \mathfrak{a}$ imply $xy \in \mathfrak{a}$). The quotient group $A/\mathfrak{a}$ inherits a uniquely defined multiplication from A which makes it into a ring, called the *quotient ring* (or residue-class ring) $A/\mathfrak{a}$. The elements of $A/\mathfrak{a}$ are the cosets of $\mathfrak{a}$ in A, and the mapping $\phi: A \to A/\mathfrak{a}$ which maps each $x \in A$ to its coset $x + \mathfrak{a}$ is a surjective ring homomorphism.

We shall frequently use the following fact:

Proposition 1.1. *There is a one-to-one order-preserving correspondence between the ideals* $\mathfrak{b}$ *of* A *which contain* $\mathfrak{a}$, *and the ideals* $\bar{\mathfrak{b}}$ *of* $A/\mathfrak{a}$, *given by* $\mathfrak{b} = \phi^{-1}(\bar{\mathfrak{b}})$. ∎

If $f: A \to B$ is any ring homomorphism, the *kernel* of f $(=f^{-1}(0))$ is an ideal $\mathfrak{a}$ of A, and the *image* of f $(=f(A))$ is a subring C of B; and f induces a ring isomorphism $A/\mathfrak{a} \cong C$.

We shall sometimes use the notation $x \equiv y \pmod{\mathfrak{a}}$; this means that $x - y \in \mathfrak{a}$.

ZERO-DIVISORS. NILPOTENT ELEMENTS. UNITS

A *zero-divisor* in a ring A is an element x which "divides 0", i.e., for which there exists $y \neq 0$ in A such that $xy = 0$. A ring with no zero-divisors $\neq 0$ (and in which $1 \neq 0$) is called an *integral domain*. For example, $\mathbf{Z}$ and $k[x_1, \ldots, x_n]$ (k a field, x_i indeterminates) are integral domains.

An element $x \in A$ is *nilpotent* if $x^n = 0$ for some $n > 0$. A nilpotent element is a zero-divisor (unless $A = 0$), but not conversely (in general).

A *unit* in A is an element x which "divides 1", i.e., an element x such that $xy = 1$ for some $y \in A$. The element y is then uniquely determined by x, and is written x^{-1}. The units in A form a (multiplicative) abelian group.

The multiples ax of an element $x \in A$ form a *principal* ideal, denoted by (x) or Ax. x is a unit $\Leftrightarrow (x) = A = (1)$. The *zero* ideal (0) is usually denoted by 0.

A *field* is a ring A in which $1 \neq 0$ and every non-zero element is a unit. Every field is an integral domain (but not conversely: **Z** is not a field).

Proposition 1.2. Let A be a ring $\neq 0$. Then the following are equivalent:

i) *A is a field;*

ii) *the only ideals in A are* 0 *and* (1)*;*

iii) *every homomorphism of A into a non-zero ring B is injective.*

Proof. i) $\Rightarrow$ ii). Let $\mathfrak{a} \neq 0$ be an ideal in A. Then $\mathfrak{a}$ contains a non-zero element x; x is a unit, hence $\mathfrak{a} \supseteq (x) = (1)$, hence $\mathfrak{a} = (1)$.

ii) $\Rightarrow$ iii). Let $\phi: A \to B$ be a ring homomorphism. Then $\text{Ker}(\phi)$ is an ideal $\neq (1)$ in A, hence $\text{Ker}(\phi) = 0$, hence ϕ is injective.

iii) $\Rightarrow$ i). Let x be an element of A which is not a unit. Then $(x) \neq (1)$, hence $B = A/(x)$ is not the zero ring. Let $\phi: A \to B$ be the natural homomorphism of A onto B, with kernel (x). By hypothesis, ϕ is injective, hence $(x) = 0$, hence $x = 0$. ∎

PRIME IDEALS AND MAXIMAL IDEALS

An ideal $\mathfrak{p}$ in A is *prime* if $\mathfrak{p} \neq (1)$ and if $xy \in \mathfrak{p} \Rightarrow x \in \mathfrak{p}$ or $y \in \mathfrak{p}$.

An ideal $\mathfrak{m}$ in A is *maximal* if $\mathfrak{m} \neq (1)$ and if there is no ideal $\mathfrak{a}$ such that $\mathfrak{m} \subset \mathfrak{a} \subset (1)$ (*strict* inclusions). Equivalently:

$\mathfrak{p}$ is prime $\Leftrightarrow A/\mathfrak{p}$ is an integral domain;

$\mathfrak{m}$ is maximal $\Leftrightarrow A/\mathfrak{m}$ is a field (by (1.1) and (1.2)).

Hence a maximal ideal is prime (but not conversely, in general). The zero ideal is prime $\Leftrightarrow A$ is an integral domain.

If $f: A \to B$ is a ring homomorphism and $\mathfrak{q}$ is a prime ideal in B, then $f^{-1}(\mathfrak{q})$ is a prime ideal in A, for $A/f^{-1}(\mathfrak{q})$ is isomorphic to a subring of $B/\mathfrak{q}$ and hence has no zero-divisor $\neq 0$. But if $\mathfrak{n}$ is a maximal ideal of B it is not necessarily true that $f^{-1}(\mathfrak{n})$ is maximal in A; all we can say for sure is that it is prime. (Example: $A = \mathbf{Z}$, $B = \mathbf{Q}$, $\mathfrak{n} = 0$.)

Prime ideals are fundamental to the whole of commutative algebra. The following theorem and its corollaries ensure that there is always a sufficient supply of them.

Theorem 1.3. Every ring $A \neq 0$ has at least one maximal ideal. (Remember that "ring" means commutative ring with 1.)

Proof. This is a standard application of Zorn's lemma.* Let Σ be the set of all ideals $\neq (1)$ in A. Order Σ by inclusion. Σ is not empty, since $0 \in \Sigma$. To apply

* Let S be a non-empty partially ordered set (i.e., we are given a relation $x \leqslant y$ on S which is reflexive and transitive and such that $x \leqslant y$ and $y \leqslant x$ together imply

Zorn's lemma we must show that every chain in Σ has an upper bound in Σ; let then $(\mathfrak{a}_\alpha)$ be a chain of ideals in Σ, so that for each pair of indices α, β we have either $\mathfrak{a}_\alpha \subseteq \mathfrak{a}_\beta$ or $\mathfrak{a}_\beta \subseteq \mathfrak{a}_\alpha$. Let $\mathfrak{a} = \bigcup_\alpha \mathfrak{a}_\alpha$. Then $\mathfrak{a}$ is an ideal (verify this) and $1 \notin \mathfrak{a}$ because $1 \notin \mathfrak{a}_\alpha$ for all α. Hence $\mathfrak{a} \in \Sigma$, and $\mathfrak{a}$ is an upper bound of the chain. Hence by Zorn's lemma Σ has a maximal element. ∎

Corollary 1.4. *If $\mathfrak{a} \neq (1)$ is an ideal of A, there exists a maximal ideal of A containing $\mathfrak{a}$.*

Proof. Apply (1.3) to $A/\mathfrak{a}$, bearing in mind (1.1). Alternatively, modify the proof of (1.3). ∎

Corollary 1.5. *Every non-unit of A is contained in a maximal ideal.* ∎

Remarks. 1) If A is Noetherian (Chapter 7) we can avoid the use of Zorn's lemma: the set of all ideals $\neq (1)$ has a maximal element.

2) There exist rings with exactly one maximal ideal, for example fields. A ring A with exactly one maximal ideal $\mathfrak{m}$ is called a *local ring*. The field $k = A/\mathfrak{m}$ is called the *residue field* of A.

Proposition 1.6. i) *Let A be a ring and $\mathfrak{m} \neq (1)$ an ideal of A such that every $x \in A - \mathfrak{m}$ is a unit in A. Then A is a local ring and $\mathfrak{m}$ its maximal ideal.*

ii) *Let A be a ring and $\mathfrak{m}$ a maximal ideal of A, such that every element of $1 + \mathfrak{m}$ (i.e., every $1 + x$, where $x \in \mathfrak{m}$) is a unit in A. Then A is a local ring.*

Proof. i) Every ideal $\neq (1)$ consists of non-units, hence is contained in $\mathfrak{m}$. Hence $\mathfrak{m}$ is the only maximal ideal of A.

ii) Let $x \in A - \mathfrak{m}$. Since $\mathfrak{m}$ is maximal, the ideal generated by x and $\mathfrak{m}$ is (1), hence there exist $y \in A$ and $t \in \mathfrak{m}$ such that $xy + t = 1$; hence $xy = 1 - t$ belongs to $1 + \mathfrak{m}$ and therefore is a unit. Now use i). ∎

A ring with only a finite number of maximal ideals is called *semi-local*.

Examples. 1) $A = k[x_1, \ldots, x_n]$, k a field. Let $f \in A$ be an irreducible polynomial. By unique factorization, the ideal (f) is prime.

2) $A = \mathbf{Z}$. Every ideal in $\mathbf{Z}$ is of the form (m) for some $m \geqslant 0$. The ideal (m) is prime $\Leftrightarrow m = 0$ or a prime number. All the ideals (p), where p is a prime number, are maximal: $\mathbf{Z}/(p)$ is the field of p elements.

The same holds in Example 1) for $n = 1$, but not for $n > 1$. The ideal $\mathfrak{m}$ of all polynomials in $A = k[x_1, \ldots, x_n]$ with zero constant term is maximal (since

---

$x = y$). A subset T of S is a *chain* if either $x \leqslant y$ or $y \leqslant x$ for every pair of elements x, y in T. Then Zorn's lemma may be stated as follows: if every chain T of S has an upper bound in S (i.e., if there exists $x \in S$ such that $t \leqslant x$ for all $t \in T$) then S has at least one maximal element.

For a proof of the equivalence of Zorn's lemma with the axiom of choice, the well-ordering principle, etc., see for example P. R. Halmos, *Naive Set Theory*, Van Nostrand (1960).

it is the kernel of the homomorphism $A \to k$ which maps $f \in A$ to $f(0)$). But if $n > 1$, $\mathfrak{m}$ is not a principal ideal: in fact it requires at least n generators.

3) A *principal ideal domain* is an integral domain in which every ideal is principal. In such a ring every non-zero prime ideal is maximal. For if $(x) \neq 0$ is a prime ideal and $(y) \supset (x)$, we have $x \in (y)$, say $x = yz$, so that $yz \in (x)$ and $y \notin (x)$, hence $z \in (x)$: say $z = tx$. Then $x = yz = ytx$, so that $yt = 1$ and therefore $(y) = (1)$.

NILRADICAL AND JACOBSON RADICAL

Proposition 1.7. *The set $\mathfrak{N}$ of all nilpotent elements in a ring A is an ideal, and $A/\mathfrak{N}$ has no nilpotent element $\neq 0$.*

Proof. If $x \in \mathfrak{N}$, clearly $ax \in \mathfrak{N}$ for all $a \in A$. Let $x, y \in \mathfrak{N}$: say $x^m = 0$, $y^n = 0$. By the binomial theorem (which is valid in any commutative ring), $(x + y)^{m+n-1}$ is a sum of integer multiples of products $x^r y^s$, where $r + s = m + n - 1$; we cannot have both $r < m$ and $s < n$, hence each of these products vanishes and therefore $(x + y)^{m+n-1} = 0$. Hence $x + y \in \mathfrak{N}$ and therefore $\mathfrak{N}$ is an ideal.

Let $\bar{x} \in A/\mathfrak{N}$ be represented by $x \in A$. Then $\bar{x}^n$ is represented by x^n, so that $\bar{x}^n = 0 \Rightarrow x^n \in \mathfrak{N} \Rightarrow (x^n)^k = 0$ for some $k > 0 \Rightarrow x \in \mathfrak{N} \Rightarrow \bar{x} = 0$. ∎

The ideal $\mathfrak{N}$ is called the *nilradical* of A. The following proposition gives an alternative definition of $\mathfrak{N}$:

Proposition 1.8. *The nilradical of A is the intersection of all the prime ideals of A.*

Proof. Let $\mathfrak{N}'$ denote the intersection of all the prime ideals of A. If $f \in A$ is nilpotent and if $\mathfrak{p}$ is a prime ideal, then $f^n = 0 \in \mathfrak{p}$ for some $n > 0$, hence $f \in \mathfrak{p}$ (because $\mathfrak{p}$ is prime). Hence $f \in \mathfrak{N}'$.

Conversely, suppose that f is not nilpotent. Let Σ be the set of ideals $\mathfrak{a}$ with the property

$$n > 0 \Rightarrow f^n \notin \mathfrak{a}.$$

Then Σ is not empty because $0 \in \Sigma$. As in (1.3) Zorn's lemma can be applied to the set Σ, ordered by inclusion, and therefore Σ has a maximal element. Let $\mathfrak{p}$ be a maximal element of Σ. We shall show that $\mathfrak{p}$ is a prime ideal. Let $x, y \notin \mathfrak{p}$. Then the ideals $\mathfrak{p} + (x)$, $\mathfrak{p} + (y)$ strictly contain $\mathfrak{p}$ and therefore do not belong to Σ; hence

$$f^m \in \mathfrak{p} + (x), \quad f^n \in \mathfrak{p} + (y)$$

for some m, n. It follows that $f^{m+n} \in \mathfrak{p} + (xy)$, hence the ideal $\mathfrak{p} + (xy)$ is not in Σ and therefore $xy \notin \mathfrak{p}$. Hence we have a prime ideal $\mathfrak{p}$ such that $f \notin \mathfrak{p}$, so that $f \notin \mathfrak{N}'$. ∎

The *Jacobson radical* $\mathfrak{R}$ of A is defined to be the intersection of all the maximal ideals of A. It can be characterized as follows:

Proposition 1.9. *$x \in \mathfrak{R} \Leftrightarrow 1 - xy$ is a unit in A for all $y \in A$.*

Proof. $\Rightarrow$: Suppose $1 - xy$ is not a unit. By (1.5) it belongs to some maximal ideal $\mathfrak{m}$; but $x \in \mathfrak{R} \subseteq \mathfrak{m}$, hence $xy \in \mathfrak{m}$ and therefore $1 \in \mathfrak{m}$, which is absurd.

$\Leftarrow$: Suppose $x \notin \mathfrak{m}$ for some maximal ideal $\mathfrak{m}$. Then $\mathfrak{m}$ and x generate the unit ideal (1), so that we have $u + xy = 1$ for some $u \in \mathfrak{m}$ and some $y \in A$. Hence $1 - xy \in \mathfrak{m}$ and is therefore not a unit. ■

OPERATIONS ON IDEALS

If $\mathfrak{a}, \mathfrak{b}$ are ideals in a ring A, their *sum* $\mathfrak{a} + \mathfrak{b}$ is the set of all $x + y$ where $x \in \mathfrak{a}$ and $y \in \mathfrak{b}$. It is the smallest ideal containing $\mathfrak{a}$ and $\mathfrak{b}$. More generally, we may define the sum $\sum_{i \in I} \mathfrak{a}_i$ of any family (possibly infinite) of ideals $\mathfrak{a}_i$ of A; its elements are all sums $\sum x_i$, where $x_i \in \mathfrak{a}_i$ for all $i \in I$ and almost all of the x_i (i.e., all but a finite set) are zero. It is the smallest ideal of A which contains all the ideals $\mathfrak{a}_i$.

The *intersection* of any family $(\mathfrak{a}_i)_{i \in I}$ of ideals is an ideal. Thus the ideals of A form a complete lattice with respect to inclusion.

The *product* of two ideals $\mathfrak{a}, \mathfrak{b}$ in A is the ideal $\mathfrak{a}\mathfrak{b}$ *generated by* all products xy, where $x \in \mathfrak{a}$ and $y \in \mathfrak{b}$. It is the set of all finite sums $\sum x_i y_i$ where each $x_i \in \mathfrak{a}$ and each $y_i \in \mathfrak{b}$. Similarly we define the product of any *finite* family of ideals. In particular the powers $\mathfrak{a}^n$ $(n > 0)$ of an ideal $\mathfrak{a}$ are defined; conventionally, $\mathfrak{a}^0 = (1)$. Thus $\mathfrak{a}^n$ $(n > 0)$ is the ideal generated by all products $x_1 x_2 \cdots x_n$ in which each factor x_i belongs to $\mathfrak{a}$.

Examples. 1) If $A = \mathbf{Z}$, $\mathfrak{a} = (m)$, $\mathfrak{b} = (n)$ then $\mathfrak{a} + \mathfrak{b}$ is the ideal generated by the h.c.f. of m and n; $\mathfrak{a} \cap \mathfrak{b}$ is the ideal generated by their l.c.m.; and $\mathfrak{a}\mathfrak{b} = (mn)$. Thus (in this case) $\mathfrak{a}\mathfrak{b} = \mathfrak{a} \cap \mathfrak{b} \Leftrightarrow m, n$ are coprime.

2) $A = k[x_1, \ldots, x_n]$, $\mathfrak{a} = (x_1, \ldots, x_n) =$ ideal generated by $x_1, \ldots, x_n$. Then $\mathfrak{a}^m$ is the set of all polynomials with no terms of degree $< m$.

The three operations so far defined (sum, intersection, product) are all commutative and associative. Also there is the *distributive law*

$$\mathfrak{a}(\mathfrak{b} + \mathfrak{c}) = \mathfrak{a}\mathfrak{b} + \mathfrak{a}\mathfrak{c}.$$

In the ring $\mathbf{Z}$, $\cap$ and $+$ are distributive over each other. This is not the case in general, and the best we have in this direction is the *modular law*

$$\mathfrak{a} \cap (\mathfrak{b} + \mathfrak{c}) = \mathfrak{a} \cap \mathfrak{b} + \mathfrak{a} \cap \mathfrak{c} \text{ if } \mathfrak{a} \supseteq \mathfrak{b} \text{ or } \mathfrak{a} \supseteq \mathfrak{c}.$$

Again, in $\mathbf{Z}$, we have $(\mathfrak{a} + \mathfrak{b})(\mathfrak{a} \cap \mathfrak{b}) = \mathfrak{a}\mathfrak{b}$; but in general we have only $(\mathfrak{a} + \mathfrak{b})(\mathfrak{a} \cap \mathfrak{b}) \subseteq \mathfrak{a}\mathfrak{b}$ (since $(\mathfrak{a} + \mathfrak{b})(\mathfrak{a} \cap \mathfrak{b}) = \mathfrak{a}(\mathfrak{a} \cap \mathfrak{b}) + \mathfrak{b}(\mathfrak{a} \cap \mathfrak{b}) \subseteq \mathfrak{a}\mathfrak{b}$). Clearly $\mathfrak{a}\mathfrak{b} \subseteq \mathfrak{a} \cap \mathfrak{b}$, hence

$$\mathfrak{a} \cap \mathfrak{b} = \mathfrak{a}\mathfrak{b} \text{ provided } \mathfrak{a} + \mathfrak{b} = (1).$$

Two ideals $\mathfrak{a}, \mathfrak{b}$ are said to be *coprime* (or comaximal) if $\mathfrak{a} + \mathfrak{b} = (1)$. Thus for coprime ideals we have $\mathfrak{a} \cap \mathfrak{b} = \mathfrak{ab}$. Clearly two ideals $\mathfrak{a}, \mathfrak{b}$ are coprime if and only if there exist $x \in \mathfrak{a}$ and $y \in \mathfrak{b}$ such that $x + y = 1$.

Let $A_1, \ldots, A_n$ be rings. Their *direct product*

$$A = \prod_{i=1}^{n} A_i$$

is the set of all sequences $x = (x_1, \ldots, x_n)$ with $x_i \in A_i$ $(1 \leqslant i \leqslant n)$ and componentwise addition and multiplication. A is a commutative ring with identity element $(1, 1, \ldots, 1)$. We have projections $p_i: A \to A_i$ defined by $p_i(x) = x_i$; they are ring homomorphisms.

Let A be a ring and $\mathfrak{a}_1, \ldots, \mathfrak{a}_n$ ideals of A. Define a homomorphism

$$\phi: A \to \prod_{i=1}^{n} (A/\mathfrak{a}_i)$$

by the rule $\phi(x) = (x + \mathfrak{a}_1, \ldots, x + \mathfrak{a}_n)$.

Proposition 1.10. i) *If $\mathfrak{a}_i, \mathfrak{a}_j$ are coprime whenever $i \neq j$, then $\Pi \mathfrak{a}_i = \bigcap \mathfrak{a}_i$.*

ii) *ϕ is surjective $\Leftrightarrow$ $\mathfrak{a}_i, \mathfrak{a}_j$ are coprime whenever $i \neq j$.*

iii) *ϕ is injective $\Leftrightarrow$ $\bigcap \mathfrak{a}_i = (0)$.*

Proof. i) by induction on n. The case $n = 2$ is dealt with above. Suppose $n > 2$ and the result true for $\mathfrak{a}_1, \ldots, \mathfrak{a}_{n-1}$, and let $\mathfrak{b} = \prod_{i=1}^{n-1} \mathfrak{a}_i = \bigcap_{i=1}^{n-1} \mathfrak{a}_i$. Since $\mathfrak{a}_i + \mathfrak{a}_n = (1)$ $(1 \leqslant i \leqslant n - 1)$ we have equations $x_i + y_i = 1$ $(x_i \in \mathfrak{a}_i, y_i \in \mathfrak{a}_n)$ and therefore

$$\prod_{i=1}^{n-1} x_i = \prod_{i=1}^{n-1} (1 - y_i) \equiv 1 \pmod{\mathfrak{a}_n}.$$

Hence $\mathfrak{a}_n + \mathfrak{b} = (1)$ and so

$$\prod_{i=1}^{n} \mathfrak{a}_i = \mathfrak{b}\mathfrak{a}_n = \mathfrak{b} \cap \mathfrak{a}_n = \bigcap_{i=1}^{n} \mathfrak{a}_i.$$

ii) $\Rightarrow$: Let us show for example that $\mathfrak{a}_1, \mathfrak{a}_2$ are coprime. There exists $x \in A$ such that $\phi(x) = (1, 0, \ldots, 0)$; hence $x \equiv 1 \pmod{\mathfrak{a}_1}$ and $x \equiv 0 \pmod{\mathfrak{a}_2}$, so that

$$1 = (1 - x) + x \in \mathfrak{a}_1 + \mathfrak{a}_2.$$

$\Leftarrow$: It is enough to show, for example, that there is an element $x \in A$ such that $\phi(x) = (1, 0, \ldots, 0)$. Since $\mathfrak{a}_1 + \mathfrak{a}_i = (1)$ $(i > 1)$ we have equations $u_i + v_i = 1$ $(u_i \in \mathfrak{a}_1, v_i \in \mathfrak{a}_i)$. Take $x = \prod_{i=2}^{n} v_i$, then $x = \Pi(1 - u_i) \equiv 1 \pmod{\mathfrak{a}_1}$, and $x \equiv 0 \pmod{\mathfrak{a}_i}$, $i > 1$. Hence $\phi(x) = (1, 0, \ldots, 0)$ as required.

iii) Clear, since $\bigcap \mathfrak{a}_i$ is the kernel of ϕ. ∎

The *union* $\mathfrak{a} \cup \mathfrak{b}$ of ideals is not in general an ideal.

Proposition 1.11. i) *Let* $\mathfrak{p}_1, \ldots, \mathfrak{p}_n$ *be prime ideals and let* $\mathfrak{a}$ *be an ideal contained in* $\bigcup_{i=1}^n \mathfrak{p}_i$. *Then* $\mathfrak{a} \subseteq \mathfrak{p}_i$ *for some* i.

ii) *Let* $\mathfrak{a}_1, \ldots, \mathfrak{a}_n$ *be ideals and let* $\mathfrak{p}$ *be a prime ideal containing* $\bigcap_{i=1}^n \mathfrak{a}_i$. *Then* $\mathfrak{p} \supseteq \mathfrak{a}_i$ *for some* i. *If* $\mathfrak{p} = \bigcap \mathfrak{a}_i$, *then* $\mathfrak{p} = \mathfrak{a}_i$ *for some* i.

Proof. i) is proved by induction on n in the form

$$\mathfrak{a} \not\subseteq \mathfrak{p}_i \ (1 \leqslant i \leqslant n) \Rightarrow \mathfrak{a} \not\subseteq \bigcup_{i=1}^n \mathfrak{p}_i.$$

It is certainly true for $n = 1$. If $n > 1$ and the result is true for $n - 1$, then for each i there exists $x_i \in \mathfrak{a}$ such that $x_i \notin \mathfrak{p}_j$ whenever $j \neq i$. If for some i we have $x_i \notin \mathfrak{p}_i$, we are through. If not, then $x_i \in \mathfrak{p}_i$ for all i. Consider the element

$$y = \sum_{i=1}^n x_1 x_2 \cdots x_{i-1} x_{i+1} x_{i+2} \cdots x_n;$$

we have $y \in \mathfrak{a}$ and $y \notin \mathfrak{p}_i$ $(1 \leqslant i \leqslant n)$. Hence $\mathfrak{a} \not\subseteq \bigcup_{i=1}^n \mathfrak{p}_i$.

ii) Suppose $\mathfrak{p} \not\supseteq \mathfrak{a}_i$ for all i. Then there exist $x_i \in \mathfrak{a}_i$, $x_i \notin \mathfrak{p}$ $(1 \leqslant i \leqslant n)$, and therefore $\Pi x_i \in \Pi \mathfrak{a}_i \subseteq \bigcap \mathfrak{a}_i$; but $\Pi x_i \notin \mathfrak{p}$ (since $\mathfrak{p}$ is prime). Hence $\mathfrak{p} \not\supseteq \bigcap \mathfrak{a}_i$. Finally, if $\mathfrak{p} = \bigcap \mathfrak{a}_i$, then $\mathfrak{p} \subseteq \mathfrak{a}_i$ and hence $\mathfrak{p} = \mathfrak{a}_i$ for some i. ■

If $\mathfrak{a}, \mathfrak{b}$ are ideals in a ring A, their *ideal quotient* is

$$(\mathfrak{a}:\mathfrak{b}) = \{x \in A : x\mathfrak{b} \subseteq \mathfrak{a}\}$$

which is an ideal. In particular, $(0:\mathfrak{b})$ is called the *annihilator* of $\mathfrak{b}$ and is also denoted by Ann $(\mathfrak{b})$: it is the set of all $x \in A$ such that $x\mathfrak{b} = 0$. In this notation the set of all zero-divisors in A is

$$D = \bigcup_{x \neq 0} \operatorname{Ann}(x).$$

If $\mathfrak{b}$ is a principal ideal (x), we shall write $(\mathfrak{a} : x)$ in place of $(\mathfrak{a} : (x))$.

Example. If $A = \mathbf{Z}$, $\mathfrak{a} = (m)$, $\mathfrak{b} = (n)$, where say $m = \prod_p p^{\mu_p}$, $n = \prod_p p^{\nu_p}$, then $(\mathfrak{a}:\mathfrak{b}) = (q)$ where $q = \prod_p p^{\gamma_p}$ and

$$\gamma_p = \max(\mu_p - \nu_p, 0) = \mu_p - \min(\mu_p, \nu_p).$$

Hence $q = m/(m, n)$, where (m, n) is the h.c.f. of m and n.

Exercise 1.12. i) $\mathfrak{a} \subseteq (\mathfrak{a}:\mathfrak{b})$

ii) $(\mathfrak{a}:\mathfrak{b})\mathfrak{b} \subseteq \mathfrak{a}$

iii) $((\mathfrak{a}:\mathfrak{b}):\mathfrak{c}) = (\mathfrak{a}:\mathfrak{bc}) = ((\mathfrak{a}:\mathfrak{c}):\mathfrak{b})$

iv) $(\bigcap_i \mathfrak{a}_i : \mathfrak{b}) = \bigcap_i (\mathfrak{a}_i : \mathfrak{b})$

v) $(\mathfrak{a} : \sum_i \mathfrak{b}_i) = \bigcap_i (\mathfrak{a} : \mathfrak{b}_i)$.

If $\mathfrak{a}$ is any ideal of A, the *radical* of $\mathfrak{a}$ is

$$r(\mathfrak{a}) = \{x \in A : x^n \in \mathfrak{a} \text{ for some } n > 0\}.$$

If $\phi: A \to A/\mathfrak{a}$ is the standard homomorphism, then $r(\mathfrak{a}) = \phi^{-1}(\mathfrak{R}_{A/\mathfrak{a}})$ and hence $r(\mathfrak{a})$ is an ideal by (1.7).

Exercise 1.13. i) $r(\mathfrak{a}) \supseteq \mathfrak{a}$

ii) $r(r(\mathfrak{a})) = r(\mathfrak{a})$

iii) $r(\mathfrak{ab}) = r(\mathfrak{a} \cap \mathfrak{b}) = r(\mathfrak{a}) \cap r(\mathfrak{b})$

iv) $r(\mathfrak{a}) = (1) \Leftrightarrow \mathfrak{a} = (1)$

v) $r(\mathfrak{a} + \mathfrak{b}) = r(r(\mathfrak{a}) + r(\mathfrak{b}))$

vi) *if* $\mathfrak{p}$ *is prime,* $r(\mathfrak{p}^n) = \mathfrak{p}$ *for all* $n > 0$.

Proposition 1.14. *The radical of an ideal* $\mathfrak{a}$ *is the intersection of the prime ideals which contain* $\mathfrak{a}$.

Proof. Apply (1.8) to $A/\mathfrak{a}$. ∎

More generally, we may define the radical $r(E)$ of any *subset* E of A in the same way. It is *not* an ideal in general. We have $r(\bigcup_\alpha E_\alpha) = \bigcup r(E_\alpha)$, for any family of subsets E_α of A.

Proposition 1.15. D = *set of zero-divisors of* $A = \bigcup_{x \neq 0} r(\text{Ann}(x))$.

Proof. $D = r(D) = r(\bigcup_{x \neq 0} \text{Ann}(x)) = \bigcup_{x \neq 0} r(\text{Ann}(x))$. ∎

Example. If $A = \mathbf{Z}$, $\mathfrak{a} = (m)$, let p_i $(1 \leqslant i \leqslant r)$ be the distinct prime divisors of m. Then $r(\mathfrak{a}) = (p_1 \cdots p_r) = \bigcap_{i=1}^{r} (p_i)$.

Proposition 1.16. *Let* $\mathfrak{a}, \mathfrak{b}$ *be ideals in a ring* A *such that* $r(\mathfrak{a}), r(\mathfrak{b})$ *are coprime. Then* $\mathfrak{a}, \mathfrak{b}$ *are coprime.*

Proof. $r(\mathfrak{a} + \mathfrak{b}) = r(r(\mathfrak{a}) + r(\mathfrak{b})) = r(1) = (1)$, hence $\mathfrak{a} + \mathfrak{b} = (1)$ by (1.13).

EXTENSION AND CONTRACTION

Let $f: A \to B$ be a ring homomorphism. If $\mathfrak{a}$ is an ideal in A, the set $f(\mathfrak{a})$ is not necessarily an ideal in B (e.g., let f be the embedding of $\mathbf{Z}$ in $\mathbf{Q}$, the field of rationals, and take $\mathfrak{a}$ to be any non-zero ideal in $\mathbf{Z}$.) We define the *extension* $\mathfrak{a}^e$ of $\mathfrak{a}$ to be the ideal $Bf(\mathfrak{a})$ generated by $f(\mathfrak{a})$ in B: explicitly, $\mathfrak{a}^e$ is the set of all sums $\sum y_i f(x_i)$ where $x_i \in \mathfrak{a}, y_i \in B$.

If $\mathfrak{b}$ is an ideal of B, then $f^{-1}(\mathfrak{b})$ is always an ideal of A, called the *contraction* $\mathfrak{b}^c$ of $\mathfrak{b}$. If $\mathfrak{b}$ is prime, then $\mathfrak{b}^c$ is prime. If $\mathfrak{a}$ is prime, $\mathfrak{a}^e$ need not be prime (for example, $f: \mathbf{Z} \to \mathbf{Q}$, $\mathfrak{a} \neq 0$; then $\mathfrak{a}^e = \mathbf{Q}$, which is not a prime ideal).

We can factorize f as follows:

$$A \xrightarrow{p} f(A) \xrightarrow{j} B$$

where p is surjective and j is injective. For p the situation is very simple (1.1): there is a one-to-one correspondence between ideals of $f(A)$ and ideals of A which contain Ker (f), and prime ideals correspond to prime ideals. For j, on the other hand, the general situation is very complicated. The classical example is from algebraic number theory.

Example. Consider $\mathbf{Z} \to \mathbf{Z}[i]$, where $i = \sqrt{-1}$. A prime ideal (p) of $\mathbf{Z}$ may or may not stay prime when extended to $\mathbf{Z}[i]$. In fact $\mathbf{Z}[i]$ is a principal ideal domain (because it has a Euclidean algorithm) and the situation is as follows:

i) $(2)^e = ((1 + i)^2)$, the *square* of a prime ideal in $\mathbf{Z}[i]$;

ii) If $p \equiv 1 \pmod 4$ then $(p)^e$ is the product of two distinct prime ideals (for example, $(5)^e = (2 + i)(2 - i)$);

iii) If $p \equiv 3 \pmod 4$ then $(p)^e$ is prime in $\mathbf{Z}[i]$.

Of these, ii) is not a trivial result. It is effectively equivalent to a theorem of Fermat which says that a prime $p \equiv 1 \pmod 4$ can be expressed, essentially uniquely, as a sum of two integer squares (thus $5 = 2^2 + 1^2, 97 = 9^2 + 4^2$, etc.).

In fact the behavior of prime ideals under extensions of this sort is one of the central problems of algebraic number theory.

Let $f: A \to B$, $\mathfrak{a}$ and $\mathfrak{b}$ be as before. Then

Proposition 1.17. i) $\mathfrak{a} \subseteq \mathfrak{a}^{ec}$, $\mathfrak{b} \supseteq \mathfrak{b}^{ce}$;

ii) $\mathfrak{b}^c = \mathfrak{b}^{cec}$, $\mathfrak{a}^e = \mathfrak{a}^{ece}$;

iii) *If C is the set of contracted ideals in A and if E is the set of extended ideals in B, then $C = \{\mathfrak{a} \mid \mathfrak{a}^{ec} = \mathfrak{a}\}$, $E = \{\mathfrak{b} \mid \mathfrak{b}^{ce} = \mathfrak{b}\}$, and $\mathfrak{a} \mapsto \mathfrak{a}^e$ is a bijective map of C onto E, whose inverse is $\mathfrak{b} \mapsto \mathfrak{b}^c$.*

Proof. i) is trivial, and ii) follows from i).

iii) If $\mathfrak{a} \in C$, then $\mathfrak{a} = \mathfrak{b}^c = \mathfrak{b}^{cec} = \mathfrak{a}^{ec}$; conversely if $\mathfrak{a} = \mathfrak{a}^{ec}$ then $\mathfrak{a}$ is the contraction of $\mathfrak{a}^e$. Similarly for E. ∎

Exercise 1.18. *If $\mathfrak{a}_1, \mathfrak{a}_2$ are ideals of A and if $\mathfrak{b}_1, \mathfrak{b}_2$ are ideals of B, then*

$$\begin{array}{ll}
(\mathfrak{a}_1 + \mathfrak{a}_2)^e = \mathfrak{a}_1^e + \mathfrak{a}_2^e, & (\mathfrak{b}_1 + \mathfrak{b}_2)^c \supseteq \mathfrak{b}_1^c + \mathfrak{b}_2^c, \\
(\mathfrak{a}_1 \cap \mathfrak{a}_2)^e \subseteq \mathfrak{a}_1^e \cap \mathfrak{a}_2^e, & (\mathfrak{b}_1 \cap \mathfrak{b}_2)^c = \mathfrak{b}_1^c \cap \mathfrak{b}_2^c, \\
(\mathfrak{a}_1\mathfrak{a}_2)^e = \mathfrak{a}_1^e\mathfrak{a}_2^e, & (\mathfrak{b}_1\mathfrak{b}_2)^c \supseteq \mathfrak{b}_1^c\mathfrak{b}_2^c, \\
(\mathfrak{a}_1 : \mathfrak{a}_2)^e \subseteq (\mathfrak{a}_1^e : \mathfrak{a}_2^e), & (\mathfrak{b}_1 : \mathfrak{b}_2)^c \subseteq (\mathfrak{b}_1^c : \mathfrak{b}_2^c), \\
r(\mathfrak{a})^e \subseteq r(\mathfrak{a}^e), & r(\mathfrak{b})^c = r(\mathfrak{b}^c).
\end{array}$$

The set of ideals E is closed under sum and product, and C is closed under the other three operations.

EXERCISES

1. Let x be a nilpotent element of a ring A. Show that $1 + x$ is a unit of A. Deduce that the sum of a nilpotent element and a unit is a unit.

2. Let A be a ring and let $A[x]$ be the ring of polynomials in an indeterminate x, with coefficients in A. Let $f = a_0 + a_1x + \cdots + a_nx^n \in A[x]$. Prove that

i) f is a unit in $A[x] \Leftrightarrow a_0$ is a unit in A and $a_1, \ldots, a_n$ are nilpotent. [If $b_0 + b_1x + \cdots + b_mx^m$ is the inverse of f, prove by induction on r that $a_n^{r+1}b_{m-r} = 0$. Hence show that a_n is nilpotent, and then use Ex. 1.]

ii) f is nilpotent $\Leftrightarrow a_0, a_1, \ldots, a_n$ are nilpotent.

iii) f is a zero-divisor $\Leftrightarrow$ there exists $a \neq 0$ in A such that $af = 0$. [Choose a polynomial $g = b_0 + b_1x + \cdots + b_mx^m$ of least degree m such that $fg = 0$. Then $a_nb_m = 0$, hence $a_ng = 0$ (because a_ng annihilates f and has degree $< m$). Now show by induction that $a_{n-r}g = 0$ $(0 \leqslant r \leqslant n)$.]

iv) f is said to be *primitive* if $(a_0, a_1, \ldots, a_n) = (1)$. Prove that if $f, g \in A[x]$, then fg is primitive $\Leftrightarrow f$ and g are primitive.

3. Generalize the results of Exercise 2 to a polynomial ring $A[x_1, \ldots, x_r]$ in several indeterminates.

4. In the ring $A[x]$, the Jacobson radical is equal to the nilradical.

5. Let A be a ring and let $A[[x]]$ be the ring of formal power series $f = \sum_{n=0}^{\infty} a_nx^n$ with coefficients in A. Show that

i) f is a unit in $A[[x]] \Leftrightarrow a_0$ is a unit in A.

ii) If f is nilpotent, then a_n is nilpotent for all $n \geqslant 0$. Is the converse true? (See Chapter 7, Exercise 2.)

iii) f belongs to the Jacobson radical of $A[[x]] \Leftrightarrow a_0$ belongs to the Jacobson radical of A.

iv) The contraction of a maximal ideal $\mathfrak{m}$ of $A[[x]]$ is a maximal ideal of A, and $\mathfrak{m}$ is generated by $\mathfrak{m}^c$ and x.

v) Every prime ideal of A is the contraction of a prime ideal of $A[[x]]$.

6. A ring A is such that every ideal not contained in the nilradical contains a non-zero idempotent (that is, an element e such that $e^2 = e \neq 0$). Prove that the nilradical and Jacobson radical of A are equal.

7. Let A be a ring in which every element x satisfies $x^n = x$ for some $n > 1$ (depending on x). Show that every prime ideal in A is maximal.

8. Let A be a ring $\neq 0$. Show that the set of prime ideals of A has minimal elements with respect to inclusion.

9. Let $\mathfrak{a}$ be an ideal $\neq (1)$ in a ring A. Show that $\mathfrak{a} = r(\mathfrak{a}) \Leftrightarrow \mathfrak{a}$ is an intersection of prime ideals.

10. Let A be a ring, $\mathfrak{R}$ its nilradical. Show that the following are equivalent:

i) A has exactly one prime ideal;

ii) every element of A is either a unit or nilpotent;

iii) $A/\mathfrak{R}$ is a field.

11. A ring A is *Boolean* if $x^2 = x$ for all $x \in A$. In a Boolean ring A, show that

i) $2x = 0$ for all $x \in A$;

ii) every prime ideal $\mathfrak{p}$ is maximal, and $A/\mathfrak{p}$ is a field with two elements;

iii) every finitely generated ideal in A is principal.

12. A local ring contains no idempotent $\neq 0, 1$.

Construction of an algebraic closure of a field (E. Artin).

13. Let K be a field and let Σ be the set of all irreducible monic polynomials f in one

indeterminate with coefficients in K. Let A be the polynomial ring over K generated by indeterminates x_f, one for each $f \in \Sigma$. Let $\mathfrak{a}$ be the ideal of A generated by the polynomials $f(x_f)$ for all $f \in \Sigma$. Show that $\mathfrak{a} \neq (1)$.

Let $\mathfrak{m}$ be a maximal ideal of A containing $\mathfrak{a}$, and let $K_1 = A/\mathfrak{m}$. Then K_1 is an extension field of K in which each $f \in \Sigma$ has a root. Repeat the construction with K_1 in place of K, obtaining a field K_2, and so on. Let $L = \bigcup_{n=1}^{\infty} K_n$. Then L is a field in which each $f \in \Sigma$ splits completely into linear factors. Let $\bar{K}$ be the set of all elements of L which are algebraic over K. Then $\bar{K}$ is an algebraic closure of K.

14. In a ring A, let Σ be the set of all ideals in which every element is a zero-divisor. Show that the set Σ has maximal elements and that every maximal element of Σ is a prime ideal. Hence the set of zero-divisors in A is a union of prime ideals.

The prime spectrum of a ring

15. Let A be a ring and let X be the set of all prime ideals of A. For each subset E of A, let $V(E)$ denote the set of all prime ideals of A which contain E. Prove that
    i) if $\mathfrak{a}$ is the ideal generated by E, then $V(E) = V(\mathfrak{a}) = V(r(\mathfrak{a}))$.
    ii) $V(0) = X$, $V(1) = \varnothing$.
    iii) if $(E_i)_{i \in I}$ is any family of subsets of A, then
$$V\Big(\bigcup_{i \in I} E_i\Big) = \bigcap_{i \in I} V(E_i).$$
    iv) $V(\mathfrak{a} \cap \mathfrak{b}) = V(\mathfrak{ab}) = V(\mathfrak{a}) \cup V(\mathfrak{b})$ for any ideals $\mathfrak{a}, \mathfrak{b}$ of A.

    These results show that the sets $V(E)$ satisfy the axioms for closed sets in a topological space. The resulting topology is called the *Zariski topology*. The topological space X is called the *prime spectrum* of A, and is written Spec (A).

16. Draw pictures of Spec ($\mathbf{Z}$), Spec ($\mathbf{R}$), Spec ($\mathbf{C}[x]$), Spec ($\mathbf{R}[x]$), Spec ($\mathbf{Z}[x]$).

17. For each $f \in A$, let X_f denote the complement of $V(f)$ in $X = \text{Spec}(A)$. The sets X_f are open. Show that they form a basis of open sets for the Zariski topology, and that
    i) $X_f \cap X_g = X_{fg}$;
    ii) $X_f = \varnothing \Leftrightarrow f$ is nilpotent;
    iii) $X_f = X \Leftrightarrow f$ is a unit;
    iv) $X_f = X_g \Leftrightarrow r((f)) = r((g))$;
    v) X is quasi-compact (that is, every open covering of X has a finite sub-covering).
    vi) More generally, each X_f is quasi-compact.
    vii) An open subset of X is quasi-compact if and only if it is a finite union of sets X_f.

    The sets X_f are called *basic open sets* of $X = \text{Spec}(A)$.

    [To prove (v), remark that it is enough to consider a covering of X by basic open sets X_{f_i} $(i \in I)$. Show that the f_i generate the unit ideal and hence that there is an equation of the form
$$1 = \sum_{i \in J} g_i f_i \qquad (g_i \in A)$$
    where J is some *finite* subset of I. Then the X_{f_i} $(i \in J)$ cover X.]

18. For psychological reasons it is sometimes convenient to denote a prime ideal of A by a letter such as x or y when thinking of it as a point of $X = \text{Spec}(A)$. When thinking of x as a prime ideal of A, we denote it by $\mathfrak{p}_x$ (logically, of course, it is the same thing). Show that

i) the set $\{x\}$ is closed (we say that x is a "closed point") in $\text{Spec}(A) \Leftrightarrow \mathfrak{p}_x$ is maximal;

ii) $\overline{\{x\}} = V(\mathfrak{p}_x)$;

iii) $y \in \overline{\{x\}} \Leftrightarrow \mathfrak{p}_x \subseteq \mathfrak{p}_y$;

iv) X is a T_0-space (this means that if x, y are distinct points of X, then either there is a neighborhood of x which does not contain y, or else there is a neighborhood of y which does not contain x).

19. A topological space X is said to be *irreducible* if $X \neq \emptyset$ and if every pair of non-empty open sets in X intersect, or equivalently if every non-empty open set is dense in X. Show that Spec (A) is irreducible if and only if the nilradical of A is a prime ideal.

20. Let X be a topological space.

i) If Y is an irreducible (Exercise 19) subspace of X, then the closure $\bar{Y}$ of Y in X is irreducible.

ii) Every irreducible subspace of X is contained in a maximal irreducible subspace.

iii) The maximal irreducible subspaces of X are closed and cover X. They are called the *irreducible components* of X. What are the irreducible components of a Hausdorff space?

iv) If A is a ring and $X = \text{Spec}(A)$, then the irreducible components of X are the closed sets $V(\mathfrak{p})$, where $\mathfrak{p}$ is a minimal prime ideal of A (Exercise 8).

21. Let $\phi: A \to B$ be a ring homomorphism. Let $X = \text{Spec}(A)$ and $Y = \text{Spec}(B)$. If $\mathfrak{q} \in Y$, then $\phi^{-1}(\mathfrak{q})$ is a prime ideal of A, i.e., a point of X. Hence ϕ induces a mapping $\phi^*: Y \to X$. Show that

i) If $f \in A$ then $\phi^{*-1}(X_f) = Y_{\phi(f)}$, and hence that ϕ^* is continuous.

ii) If $\mathfrak{a}$ is an ideal of A, then $\phi^{*-1}(V(\mathfrak{a})) = V(\mathfrak{a}^e)$.

iii) If $\mathfrak{b}$ is an ideal of B, then $\overline{\phi^*(V(\mathfrak{b}))} = V(\mathfrak{b}^c)$.

iv) If ϕ is surjective, then ϕ^* is a homeomorphism of Y onto the closed subset $V(\text{Ker}(\phi))$ of X. (In particular, Spec (A) and Spec $(A/\mathfrak{R})$ (where $\mathfrak{R}$ is the nilradical of A) are naturally homeomorphic.)

v) If ϕ is injective, then $\phi^*(Y)$ is dense in X. More precisely, $\phi^*(Y)$ is dense in $X \Leftrightarrow \text{Ker}(\phi) \subseteq \mathfrak{R}$.

vi) Let $\psi: B \to C$ be another ring homomorphism. Then $(\psi \circ \phi)^* = \phi^* \circ \psi^*$.

vii) Let A be an integral domain with just one non-zero prime ideal $\mathfrak{p}$, and let K be the field of fractions of A. Let $B = (A/\mathfrak{p}) \times K$. Define $\phi: A \to B$ by $\phi(x) = (\bar{x}, x)$, where $\bar{x}$ is the image of x in $A/\mathfrak{p}$. Show that ϕ^* is bijective but not a homeomorphism.

22. Let $A = \prod_{i=1}^{n} A_i$ be the direct product of rings A_i. Show that Spec (A) is the disjoint union of open (and closed) subspaces X_i, where X_i is canonically homeomorphic with Spec (A_i).

Conversely, let A be any ring. Show that the following statements are equivalent:

i) $X = \text{Spec}(A)$ is disconnected.

ii) $A \cong A_1 \times A_2$ where neither of the rings A_1, A_2 is the zero ring.

iii) A contains an idempotent $\neq 0, 1$.

In particular, the spectrum of a local ring is always connected (Exercise 12).

23. Let A be a Boolean ring (Exercise 11), and let $X = \text{Spec}(A)$.

i) For each $f \in A$, the set X_f (Exercise 17) is both open and closed in X.

ii) Let $f_1, \ldots, f_n \in A$. Show that $X_{f_1} \cup \cdots \cup X_{f_n} = X_f$ for some $f \in A$.

iii) The sets X_f are the only subsets of X which are both open and closed. [Let $Y \subseteq X$ be both open and closed. Since Y is open, it is a union of basic open sets X_f. Since Y is closed and X is quasi-compact (Exercise 17), Y is quasi-compact. Hence Y is a finite union of basic open sets; now use (ii) above.]

iv) X is a compact Hausdorff space.

24. Let L be a lattice, in which the sup and inf of two elements a, b are denoted by $a \vee b$ and $a \wedge b$ respectively. L is a *Boolean lattice* (or *Boolean algebra*) if

i) L has a least element and a greatest element (denoted by 0, 1 respectively).

ii) Each of $\vee$, $\wedge$ is distributive over the other.

iii) Each $a \in L$ has a unique "complement" $a' \in L$ such that $a \vee a' = 1$ and $a \wedge a' = 0$.

(For example, the set of all subsets of a set, ordered by inclusion, is a Boolean lattice.)

Let L be a Boolean lattice. Define addition and multiplication in L by the rules

$$a + b = (a \wedge b') \vee (a' \wedge b), \qquad ab = a \wedge b.$$

Verify that in this way L becomes a Boolean ring, say $A(L)$.

Conversely, starting from a Boolean ring A, define an ordering on A as follows: $a \leqslant b$ means that $a = ab$. Show that, with respect to this ordering, A is a Boolean lattice. [The sup and inf are given by $a \vee b = a + b + ab$ and $a \wedge b = ab$, and the complement by $a' = 1 - a$.] In this way we obtain a one-to-one correspondence between (isomorphism classes of) Boolean rings and (isomorphism classes of) Boolean lattices.

25. From the last two exercises deduce Stone's theorem, that every Boolean lattice is isomorphic to the lattice of open-and-closed subsets of some compact Hausdorff topological space.

26. Let A be a ring. The subspace of Spec (A) consisting of the *maximal* ideals of A, with the induced topology, is called the *maximal spectrum* of A and is denoted by Max (A). For arbitrary commutative rings it does not have the nice functorial properties of Spec (A) (see Exercise 21), because the inverse image of a maximal ideal under a ring homomorphism need not be maximal.

Let X be a compact Hausdorff space and let $C(X)$ denote the ring of all real-valued continuous functions on X (add and multiply functions by adding and multiplying their values). For each $x \in X$, let $\mathfrak{m}_x$ be the set of all $f \in C(X)$ such that $f(x) = 0$. The ideal $\mathfrak{m}_x$ is maximal, because it is the kernel of the (surjective) homomorphism $C(X) \to \mathbf{R}$ which takes f to $f(x)$. If $\tilde{X}$ denotes $\operatorname{Max}(C(X))$, we have therefore defined a mapping $\mu: X \to \tilde{X}$, namely $x \mapsto \mathfrak{m}_x$.

We shall show that μ is a homeomorphism of X onto $\tilde{X}$.

i) Let $\mathfrak{m}$ be any maximal ideal of $C(X)$, and let $V = V(\mathfrak{m})$ be the set of common zeros of the functions in $\mathfrak{m}$: that is,

$$V = \{x \in X : f(x) = 0 \text{ for all } f \in \mathfrak{m}\}.$$

Suppose that V is empty. Then for each $x \in X$ there exists $f_x \in \mathfrak{m}$ such that $f_x(x) \neq 0$. Since f_x is continuous, there is an open neighborhood U_x of x in X on which f_x does not vanish. By compactness a finite number of the neighborhoods, say $U_{x_1}, \ldots, U_{x_n}$, cover X. Let

$$f = f_{x_1}^2 + \cdots + f_{x_n}^2.$$

Then f does not vanish at any point of X, hence is a unit in $C(X)$. But this contradicts $f \in \mathfrak{m}$, hence V is not empty.

Let x be a point of V. Then $\mathfrak{m} \subseteq \mathfrak{m}_x$, hence $\mathfrak{m} = \mathfrak{m}_x$ because $\mathfrak{m}$ is maximal. Hence μ is surjective.

ii) By Urysohn's lemma (this is the only non-trivial fact required in the argument) the continuous functions separate the points of X. Hence $x \neq y \Rightarrow \mathfrak{m}_x \neq \mathfrak{m}_y$, and therefore μ is injective.

iii) Let $f \in C(X)$; let

$$U_f = \{x \in X : f(x) \neq 0\}$$

and let

$$\tilde{U}_f = \{\mathfrak{m} \in \tilde{X} : f \notin \mathfrak{m}\}$$

Show that $\mu(U_f) = \tilde{U}_f$. The open sets U_f (resp. $\tilde{U}_f$) form a basis of the topology of X (resp. $\tilde{X}$) and therefore μ is a homeomorphism.

Thus X can be reconstructed from the ring of functions $C(X)$.

Affine algebraic varieties

27. Let k be an algebraically closed field and let

$$f_\alpha(t_1, \ldots, t_n) = 0$$

be a set of polynomial equations in n variables with coefficients in k. The set X of all points $x = (x_1, \ldots, x_n) \in k^n$ which satisfy these equations is an *affine algebraic variety*.

Consider the set of all polynomials $g \in k[t_1, \ldots, t_n]$ with the property that $g(x) = 0$ for all $x \in X$. This set is an ideal $I(X)$ in the polynomial ring, and is called the *ideal of the variety* X. The quotient ring

$$P(X) = k[t_1, \ldots, t_n]/I(X)$$

is the ring of polynomial functions on X, because two polynomials g, h define the same polynomial function on X if and only if $g - h$ vanishes at every point of X, that is, if and only if $g - h \in I(X)$.

Let ξ_i be the image of t_i in $P(X)$. The ξ_i $(1 \leq i \leq n)$ are the *coordinate functions* on X: if $x \in X$, then $\xi_i(x)$ is the ith coordinate of x. $P(X)$ is generated as a k-algebra by the coordinate functions, and is called the *coordinate ring* (or affine algebra) of X.

As in Exercise 26, for each $x \in X$ let $\mathfrak{m}_x$ be the ideal of all $f \in P(X)$ such that $f(x) = 0$; it is a maximal ideal of $P(X)$. Hence, if $\tilde{X} = \text{Max}\ (P(X))$, we have defined a mapping $\mu: X \to \tilde{X}$, namely $x \mapsto \mathfrak{m}_x$.

It is easy to show that μ is injective: if $x \neq y$, we must have $x_i \neq y_i$ for for some i $(1 \leq i \leq n)$, and hence $\xi_i - x_i$ is in $\mathfrak{m}_x$ but not in $\mathfrak{m}_y$, so that $\mathfrak{m}_x \neq \mathfrak{m}_y$. What is less obvious (but still true) is that μ is *surjective*. This is one form of Hilbert's Nullstellensatz (see Chapter 7).

28. Let $f_1, \ldots, f_m$ be elements of $k[t_1, \ldots, t_n]$. They determine a *polynomial mapping* $\phi: k^n \to k^m$: if $x \in k^n$, the coordinates of $\phi(x)$ are $f_1(x), \ldots, f_m(x)$.

Let X, Y be affine algebraic varieties in k^n, k^m respectively. A mapping $\phi: X \to Y$ is said to be *regular* if ϕ is the restriction to X of a polynomial mapping from k^n to k^m.

If η is a polynomial function on Y, then $\eta \circ \phi$ is a polynomial function on X. Hence ϕ induces a k-algebra homomorphism $P(Y) \to P(X)$, namely $\eta \mapsto \eta \circ \phi$. Show that in this way we obtain a one-to-one correspondence between the regular mappings $X \to Y$ and the k-algebra homomorphisms $P(Y) \to P(X)$.

2

Modules

One of the things which distinguishes the modern approach to Commutative Algebra is the greater emphasis on modules, rather than just on ideals. The extra "elbow-room" that this gives makes for greater clarity and simplicity. For instance, an ideal $\mathfrak{a}$ and its quotient ring $A/\mathfrak{a}$ are both examples of modules and so, to a certain extent, can be treated on an equal footing. In this chapter we give the definition and elementary properties of modules. We also give a brief treatment of tensor products, including a discussion of how they behave for exact sequences.

MODULES AND MODULE HOMOMORPHISMS

Let A be a ring (commutative, as always). An *A-module* is an abelian group M (written additively) on which A acts linearly: more precisely, it is a pair (M, μ), where M is an abelian group and μ is a mapping of $A \times M$ into M such that, if we write ax for $\mu(a, x)$ $(a \in A, x \in M)$, the following axioms are satisfied:

$$\begin{aligned} a(x + y) &= ax + ay, \\ (a + b)x &= ax + bx, \\ (ab)x &= a(bx), \\ 1x &= x \qquad\qquad (a, b \in A; \quad x, y \in M). \end{aligned}$$

(Equivalently, M is an abelian group together with a ring homomorphism $A \to E(M)$, where $E(M)$ is the ring of endomorphisms of the abelian group M.)

The notion of a module is a common generalization of several familiar concepts, as the following examples show:

Examples. 1) An ideal $\mathfrak{a}$ of A is an A-module. In particular A itself is an A-module.

2) If A is a field k, then A-module $=$ k-vector space.

3) $A = \mathbf{Z}$, then $\mathbf{Z}$-module $=$ abelian group (define nx to be $x + \cdots + x$).

4) $A = k[x]$ where k is a field; an A-module is a k-vector space with a linear transformation.

5) G = finite group, $A = k[G]$ = group-algebra of G over the field k (thus A is not commutative, unless G is). Then A-module $=$ k-representation of G.

Let M, N be A-modules. A mapping $f: M \to N$ is an *A-module homomorphism* (or is *A-linear*) if

$$f(x + y) = f(x) + f(y)$$
$$f(ax) = a \cdot f(x)$$

for all $a \in A$ and all $x, y \in M$. Thus f is a homomorphism of abelian groups which commutes with the action of each $a \in A$. If A is a field, an A-module homomorphism is the same thing as a linear transformation of vector spaces.

The composition of A-module homomorphisms is again an A-module homomorphism.

The set of all A-module homomorphisms from M to N can be turned into an A-module as follows: we define $f + g$ and af by the rules

$$(f + g)(x) = f(x) + g(x),$$
$$(af)(x) = a \cdot f(x)$$

for all $x \in M$. It is a trivial matter to check that the axioms for an A-module are satisfied. This A-module is denoted by $\mathrm{Hom}_A(M, N)$ (or just $\mathrm{Hom}(M, N)$ if there is no ambiguity about the ring A).

Homomorphisms $u: M' \to M$ and $v: N \to N''$ induce mappings

$$\bar{u}: \mathrm{Hom}(M, N) \to \mathrm{Hom}(M', N) \quad \text{and} \quad \bar{v}: \mathrm{Hom}(M, N) \to \mathrm{Hom}(M, N'')$$

defined as follows:

$$\bar{u}(f) = f \circ u, \quad \bar{v}(f) = v \circ f.$$

These mappings are A-module homomorphisms.

For any module M there is a natural isomorphism $\mathrm{Hom}(A, M) \cong M$: any A-module homomorphism $f: A \to M$ is uniquely determined by $f(1)$, which can be any element of M.

SUBMODULES AND QUOTIENT MODULES

A *submodule* M' of M is a subgroup of M which is closed under multiplication by elements of A. The abelian group M/M' then inherits an A-module structure from M, defined by $a(x + M') = ax + M'$. The A-module M/M' is the *quotient* of M by M'. The natural map of M onto M/M' is an A-module homomorphism. There is a one-to-one order-preserving correspondence between submodules of M which contain M', and submodules of M'' (just as for ideals; the statement for ideals is a special case).

If $f: M \to N$ is an A-module homomorphism, the *kernel* of f is the set

$$\mathrm{Ker}(f) = \{x \in M : f(x) = 0\}$$

and is a submodule of M. The *image* of f is the set

$$\mathrm{Im}(f) = f(M)$$

and is a submodule of N. The *cokernel* of f is

$$\text{Coker}(f) = N/\text{Im}(f)$$

which is a quotient module of N.

If M' is a submodule of M such that $M' \subseteq \text{Ker}(f)$, then f gives rise to a homomorphism $\bar{f}: M/M' \to N$, defined as follows: if $\bar{x} \in M/M'$ is the image of $x \in M$, then $\bar{f}(\bar{x}) = f(x)$. The kernel of $\bar{f}$ is $\text{Ker}(f)/M'$. The homomorphism $\bar{f}$ is said to be *induced* by f. In particular, taking $M' = \text{Ker}(f)$, we have an isomorphism of A-modules

$$M/\text{Ker}(f) \cong \text{Im}(f).$$

OPERATIONS ON SUBMODULES

Most of the operations on ideals considered in Chapter 1 have their counterparts for modules. Let M be an A-module and let $(M_i)_{i \in I}$ be a family of submodules of M. Their *sum* $\sum M_i$ is the set of all (finite) sums $\sum x_i$, where $x_i \in M_i$ for all $i \in I$, and almost all the x_i (that is, all but a finite number) are zero. $\sum M_i$ is the smallest submodule of M which contains all the M_i.

The intersection $\bigcap M_i$ is again a submodule of M. Thus the submodules of M form a complete lattice with respect to inclusion.

Proposition 2.1. i) *If $L \supseteq M \supseteq N$ are A-modules, then*

$$(L/N)/(M/N) \cong L/M.$$

ii) *If M_1, M_2 are submodules of M, then*

$$(M_1 + M_2)/M_1 \cong M_2/(M_1 \cap M_2).$$

Proof. i) Define $\theta: L/N \to L/M$ by $\theta(x + N) = x + M$. Then θ is a well-defined A-module homomorphism of L/N onto L/M, and its kernel is M/N; hence (i).

ii) The composite homomorphism $M_2 \to M_1 + M_2 \to (M_1 + M_2)/M_1$ is surjective, and its kernel is $M_1 \cap M_2$; hence (ii). ∎

We cannot in general define the *product* of two submodules, but we can define the product $\mathfrak{a}M$, where $\mathfrak{a}$ is an ideal and M an A-module; it is the set of all finite sums $\sum a_i x_i$ with $a_i \in \mathfrak{a}$, $x_i \in M$, and is a submodule of M.

If N, P are submodules of M, we define $(N:P)$ to be the set of all $a \in A$ such that $aP \subseteq N$; it is an *ideal* of A. In particular, $(0:M)$ is the set of all $a \in A$ such that $aM = 0$; this ideal is called the *annihilator* of M and is also denoted by $\text{Ann}(M)$. If $\mathfrak{a} \subseteq \text{Ann}(M)$, we may regard M as an $A/\mathfrak{a}$-module, as follows: if $\bar{x} \in A/\mathfrak{a}$ is represented by $x \in A$, define $\bar{x}m$ to be $xm (m \in M)$: this is independent of the choice of the representative x of $\bar{x}$, since $\mathfrak{a}M = 0$.

An A-module is *faithful* if $\mathrm{Ann}(M)=0$. If $\mathrm{Ann}(M) = \mathfrak{a}$, then M is faithful as an $A/\mathfrak{a}$-module.

Exercise 2.2. i) $\mathrm{Ann}(M + N) = \mathrm{Ann}(M) \cap \mathrm{Ann}(N)$.

ii) $(N:P) = \mathrm{Ann}((N + P)/N)$.

If x is an element of M, the set of all multiples $ax (a \in A)$ is a submodule of M, denoted by Ax or (x). If $M = \sum_{i \in I} Ax_i$, the x_i are said to be a *set of generators* of M; this means that every element of M can be expressed (not necessarily uniquely) as a finite linear combination of the x_i with coefficients in A. An A-module M is said to be *finitely generated* if it has a finite set of generators.

DIRECT SUM AND PRODUCT

If M, N are A-modules, their *direct sum* $M \oplus N$ is the set of all pairs (x, y) with $x \in M$, $y \in N$. This is an A-module if we define addition and scalar multiplication in the obvious way:

$$\begin{aligned}(x_1, y_1) + (x_2, y_2) &= (x_1 + x_2, y_1 + y_2)\\ a(x, y) &= (ax, ay).\end{aligned}$$

More generally, if $(M_i)_{i \in I}$ is any family of A-modules, we can define their *direct sum* $\bigoplus_{i \in I} M_i$; its elements are families $(x_i)_{i \in I}$ such that $x_i \in M_i$ for each $i \in I$ and almost all x_i are 0. If we drop the restriction on the number of non-zero x's we have the *direct product* $\prod_{i \in I} M_i$. Direct sum and direct product are therefore the same if the index set I is finite, but not otherwise, in general.

Suppose that the ring A is a direct product $\prod_{i=1}^n A_i$ (Chapter 1). Then the set of all elements of A of the form

$$(0, \ldots, 0, a_i, 0, \ldots, 0)$$

with $a_i \in A_i$ is an *ideal* $\mathfrak{a}_i$ of A (it is *not* a subring of A—except in trivial cases—because it does not contain the identity element of A). The ring A, considered as an A-module, is the direct sum of the ideals $\mathfrak{a}_1, \ldots, \mathfrak{a}_n$. Conversely, given a module decomposition

$$A = \mathfrak{a}_1 \oplus \cdots \oplus \mathfrak{a}_n$$

of A as a direct sum of ideals, we have

$$A \cong \prod_{i=1}^{n} (A/\mathfrak{b}_i)$$

where $\mathfrak{b}_i = \bigoplus_{j \neq i} \mathfrak{a}_j$. Each ideal $\mathfrak{a}_i$ is a ring (isomorphic to $A/\mathfrak{b}_i$). The identity element e_i of $\mathfrak{a}_i$ is an idempotent in A, and $\mathfrak{a}_i = (e_i)$.

FINITELY GENERATED MODULES

A *free* A-module is one which is isomorphic to an A-module of the form $\bigoplus_{i \in I} M_i$, where each $M_i \cong A$ (as an A-module). The notation $A^{(I)}$ is sometimes used. A finitely generated free A-module is therefore isomorphic to $A \oplus \cdots \oplus A$ (n summands), which is denoted by A^n. (Conventionally, A^0 is the zero module, denoted by 0.)

Proposition 2.3. *M is a finitely generated A-module $\Leftrightarrow$ M is isomorphic to a quotient of A^n for some integer $n > 0$.*

Proof. $\Rightarrow$: Let $x_1, \ldots, x_n$ generate M. Define $\phi: A^n \to M$ by $\phi(a_1, \ldots, a_n) = a_1x_1 + \cdots + a_nx_n$. Then ϕ is an A-module homomorphism *onto* M, and therefore $M \cong A^n/\text{Ker}(\phi)$.

$\Leftarrow$: We have an A-module homomorphism ϕ of A^n onto M. If $e_i = (0, \ldots, 0, 1, 0, \ldots, 0)$ (the 1 being in the ith place), then the e_i $(1 \leqslant i \leqslant n)$ generate A^n, hence the $\phi(e_i)$ generate M. ∎

Proposition 2.4. *Let M be a finitely generated A-module, let $\mathfrak{a}$ be an ideal of A, and let ϕ be an A-module endomorphism of M such that $\phi(M) \subseteq \mathfrak{a}M$. Then ϕ satisfies an equation of the form*

$$\phi^n + a_1\phi^{n-1} + \cdots + a_n = 0$$

where the a_i are in $\mathfrak{a}$.

Proof. Let $x_1, \ldots, x_n$ be a set of generators of M. Then each $\phi(x_i) \in \mathfrak{a}M$, so that we have say $\phi(x_i) = \sum_{j=1}^n a_{ij}x_j$ $(1 \leqslant i \leqslant n;\ a_{ij} \in \mathfrak{a})$, i.e.,

$$\sum_{j=1}^{n} (\delta_{ij}\phi - a_{ij})x_j = 0$$

where δ_{ij} is the Kronecker delta. By multiplying on the left by the adjoint of the matrix $(\delta_{ij}\phi - a_{ij})$ it follows that $\det(\delta_{ij}\phi - a_{ij})$ annihilates each x_i, hence is the zero endomorphism of M. Expanding out the determinant, we have an equation of the required form. ∎

Corollary 2.5. *Let M be a finitely generated A-module and let $\mathfrak{a}$ be an ideal of A such that $\mathfrak{a}M = M$. Then there exists $x \equiv 1 (\text{mod } \mathfrak{a})$ such that $xM = 0$.*

Proof. Take ϕ = identity, $x = 1 + a_1 + \cdots + a_n$ in (2.4). ∎

Proposition 2.6. (*Nakayama's lemma*). *Let M be a finitely generated A-module and $\mathfrak{a}$ an ideal of A contained in the Jacobson radical $\mathfrak{R}$ of A. Then $\mathfrak{a}M = M$ implies $M = 0$.*

First Proof. By (2.5) we have $xM = 0$ for some $x \equiv 1 \ (\text{mod } \mathfrak{R})$. By (1.9) x is a unit in A, hence $M = x^{-1}xM = 0$. ∎

Second Proof. Suppose $M \neq 0$, and let $u_1, \ldots, u_n$ be a minimal set of generators of M. Then $u_n \in \mathfrak{a}M$, hence we have an equation of the form $u_n = a_1u_1 + \cdots + a_nu_n$, with the $a_i \in \mathfrak{a}$. Hence

$$(1 - a_n)u_n = a_1u_1 + \cdots + a_{n-1}u_{n-1};$$

since $a_n \in \mathfrak{R}$, it follows from (1.9) that $1 - a_n$ is a unit in A. Hence u_n belongs to the submodule of M generated by $u_1, \ldots, u_{n-1}$: contradiction. ∎

Corollary 2.7. *Let M be a finitely generated A-module, N a submodule of M, $\mathfrak{a} \subseteq \mathfrak{R}$ an ideal. Then $M = \mathfrak{a}M + N \Rightarrow M = N$.*

Proof. Apply (2.6) to M/N, observing that $\mathfrak{a}(M/N) = (\mathfrak{a}M + N)/N$. ∎

Let A be a local ring, $\mathfrak{m}$ its maximal ideal, $k = A/\mathfrak{m}$ its residue field. Let M be a finitely generated A-module. $M/\mathfrak{m}M$ is annihilated by $\mathfrak{m}$, hence is naturally an $A/\mathfrak{m}$-module, i.e., a k-vector space, and as such is finite-dimensional.

Proposition 2.8. *Let x_i $(1 \leqslant i \leqslant n)$ be elements of M whose images in $M/\mathfrak{m}M$ form a basis of this vector space. Then the x_i generate M.*

Proof. Let N be the submodule of M generated by the x_i. Then the composite map $N \to M \to M/\mathfrak{m}M$ maps N onto $M/\mathfrak{m}M$, hence $N + \mathfrak{m}M = M$, hence $N = M$ by (2.7). ∎

EXACT SEQUENCES

A sequence of A-modules and A-homomorphisms

$$\cdots \longrightarrow M_{i-1} \xrightarrow{f_i} M_i \xrightarrow{f_{i+1}} M_{i+1} \longrightarrow \cdots \qquad (0)$$

is said to be *exact at* M_i if $\operatorname{Im}(f_i) = \operatorname{Ker}(f_{i+1})$. The sequence is *exact* if it is exact at each M_i. In particular:

$0 \to M' \xrightarrow{f} M$ is exact $\Leftrightarrow f$ is injective; (1)

$M \xrightarrow{g} M'' \to 0$ is exact $\Leftrightarrow g$ is surjective; (2)

$0 \to M' \xrightarrow{f} M \xrightarrow{g} M'' \to 0$ is exact $\Leftrightarrow f$ is injective, g is surjective and g induces an isomorphism of $\operatorname{Coker}(f) = M/f(M')$ onto M''. (3)

A sequence of type (3) is called a *short exact sequence*. Any long exact sequence (0) can be split up into short exact sequences: if $N_i = \operatorname{Im}(f_i) = \operatorname{Ker}(f_{i+1})$, we have short exact sequences $0 \to N_i \to M_i \to N_{i+1} \to 0$ for each i.

Proposition 2.9. i) *Let*

$$M' \xrightarrow{u} M \xrightarrow{v} M'' \to 0 \qquad (4)$$

be a sequence of A-modules and homomorphisms. Then the sequence (4) *is exact $\Leftrightarrow$ for all A-modules N, the sequence*

$$0 \to \operatorname{Hom}(M'', N) \xrightarrow{\bar{v}} \operatorname{Hom}(M, N) \xrightarrow{\bar{u}} \operatorname{Hom}(M', N) \qquad (4')$$

is exact.

ii) *Let*

$$0 \to N' \xrightarrow{u} N \xrightarrow{v} N'' \qquad (5)$$

be a sequence of A-modules and homomorphisms. Then the sequence (5) *is exact* $\Leftrightarrow$ *for all A-modules M, the sequence*

$$0 \to \text{Hom}(M, N') \xrightarrow{\bar{u}} \text{Hom}(M, N) \xrightarrow{\bar{v}} \text{Hom}(M, N'') \qquad (5')$$

is exact.

All four parts of this proposition are easy exercises. For example, suppose that (4′) is exact for all N. First of all, since $\bar{v}$ is injective for all N it follows that v is surjective. Next, we have $\bar{u} \circ \bar{v} = 0$, that is $v \circ u \circ f = 0$ for all $f: M'' \to N$. Taking N to be M'' and f to be the identity mapping, it follows that $v \circ u = 0$, hence $\text{Im}(u) \subseteq \text{Ker}(v)$. Next take $N = M/\text{Im}(u)$ and let $\phi: M \to N$ be the projection. Then $\phi \in \text{Ker}(\bar{u})$, hence there exists $\psi: M'' \to N$ such that $\phi = \psi \circ v$. Consequently $\text{Im}(u) = \text{Ker}(\phi) \supseteq \text{Ker}(v)$. ∎

Proposition 2.10. *Let*

$$\begin{array}{ccccccccc} 0 & \to & M' & \xrightarrow{u} & M & \xrightarrow{v} & M'' & \to & 0 \\ & & \downarrow{f'} & & \downarrow{f} & & \downarrow{f''} & & \\ 0 & \to & N' & \xrightarrow[u']{} & N & \xrightarrow[v']{} & N'' & \to & 0 \end{array}$$

be a commutative diagram of A-modules and homomorphisms, with the rows exact. Then there exists an exact sequence

$$0 \to \text{Ker}(f') \xrightarrow{\bar{u}} \text{Ker}(f) \xrightarrow{\bar{v}} \text{Ker}(f'') \xrightarrow{d} \text{Coker}(f') \xrightarrow{\bar{u}'} \text{Coker}(f) \xrightarrow{\bar{v}'} \text{Coker}(f'') \to 0 \qquad (6)$$

in which $\bar{u}$, $\bar{v}$ *are restrictions of* u, v, *and* $\bar{u}'$, $\bar{v}'$ *are induced by* u', v'.

The *boundary homomorphism* d is defined as follows: if $x'' \in \text{Ker}(f'')$, we have $x'' = v(x)$ for some $x \in M$, and $v'(f(x)) = f''(v(x)) = 0$, hence $f(x) \in \text{Ker}(v') = \text{Im}(u')$, so that $f(x) = u'(y')$ for some $y' \in N'$. Then $d(x'')$ is defined to be the image of y' in $\text{Coker}(f')$. The verification that d is well-defined, and that the sequence (6) is exact, is a straightforward exercise in diagram-chasing which we leave to the reader. ∎

Remark. (2.10) is a special case of the exact homology sequence of homological algebra.

Let C be a class of A-modules and let λ be a function on C with values in $\mathbf{Z}$ (or, more generally, with values in an abelian group G). The function λ is *additive* if, for each short exact sequence (3) in which all the terms belong to C, we have $\lambda(M') - \lambda(M) + \lambda(M'') = 0$.

Example. Let A be a field k, and let C be the class of all finite-dimensional k-vector spaces V. Then $V \mapsto \dim V$ is an additive function on C.

Proposition 2.11. Let $0 \to M_0 \to M_1 \to \cdots \to M_n \to 0$ *be an exact sequence of A-modules in which all the modules* M_i *and the kernels of all the homomorphisms belong to C. Then for any additive function* λ *on C we have*

$$\sum_{i=0}^{n} (-1)^i \lambda(M_i) = 0.$$

Proof. Split up the sequence into short exact sequences

$$0 \to N_i \to M_i \to N_{i+1} \to 0$$

($N_0 = N_{n+1} = 0$). Then we have $\lambda(M_i) = \lambda(N_i) + \lambda(N_{i+1})$. Now take the alternating sum of the $\lambda(M_i)$, and everything cancels out. ∎

TENSOR PRODUCT OF MODULES

Let M, N, P be three A-modules. A mapping $f: M \times N \to P$ is said to be *A-bilinear* if for each $x \in M$ the mapping $y \mapsto f(x, y)$ of N into P is A-linear, and for each $y \in N$ the mapping $x \mapsto f(x, y)$ of M into P is A-linear.

We shall construct an A-module T, called the *tensor product* of M and N, with the property that the A-bilinear mappings $M \times N \to P$ are in a natural one-to-one correspondence with the A-linear mappings $T \to P$, for all A-modules P. More precisely:

Proposition 2.12. Let M, N be A-modules. Then there exists a pair (T, g) consisting of an A-module T and an A-bilinear mapping $g: M \times N \to T$, with the following property:

Given any A-module P and any A-bilinear mapping $f: M \times N \to P$, there exists a unique A-linear mapping $f': T \to P$ such that $f = f' \circ g$ (in other words, every bilinear function on $M \times N$ factors through T).

Moreover, if (T, g) and (T', g') are two pairs with this property, then there exists a unique isomorphism $j: T \to T'$ such that $j \circ g = g'$.

Proof. i) *Uniqueness.* Replacing (P, f) by (T', g') we get a unique $j: T \to T'$ such that $g' = j \circ g$. Interchanging the roles of T and T', we get $j': T' \to T$ such that $g = j' \circ g'$. Each of the compositions $j \circ j', j' \circ j$ must be the identity, and therefore j is an isomorphism.

ii) *Existence.* Let C denote the free A-module $A^{(M \times N)}$. The elements of C are formal linear combinations of elements of $M \times N$ with coefficients in A, i.e. they are expressions of the form $\sum_{i=1}^{n} a_i \cdot (x_i, y_i)(a_i \in A, x_i \in M, y_i \in N)$.

Let D be the submodule of C generated by all elements of C of the following types:

$$(x + x', y) - (x, y) - (x', y)$$
$$(x, y + y') - (x, y) - (x, y')$$
$$(ax, y) - a \cdot (x, y)$$
$$(x, ay) - a \cdot (x, y).$$

Let $T = C/D$. For each basis element (x, y) of C, let $x \otimes y$ denote its image in T. Then T is generated by the elements of the form $x \otimes y$, and from our definitions we have

$$(x + x') \otimes y = x \otimes y + x' \otimes y, \quad x \otimes (y + y') = x \otimes y + x \otimes y',$$
$$(ax) \otimes y = x \otimes (ay) = a(x \otimes y)$$

Equivalently, the mapping $g: M \times N \to T$ defined by $g(x, y) = x \otimes y$ is A-bilinear.

Any map f of $M \times N$ into an A-module P extends by linearity to an A-module homomorphism $\bar{f}: C \to P$. Suppose in particular that f is A-bilinear. Then, from the definitions, $\bar{f}$ vanishes on all the generators of D, hence on the whole of D, and therefore induces a well-defined A-homomorphism f' of $T = C/D$ into P such that $f'(x \otimes y) = f(x, y)$. The mapping f' is uniquely defined by this condition, and therefore the pair (T, g) satisfy the conditions of the proposition. ∎

Remarks. i) The module T constructed above is called the *tensor product* of M and N, and is denoted by $M \otimes_A N$, or just $M \otimes N$ if there is no ambiguity about the ring A. It is generated as an A-module by the "products" $x \otimes y$. If $(x_i)_{i \in I}$, $(y_j)_{j \in J}$ are families of generators of M, N respectively, then the elements $x_i \otimes y_j$ generate $M \otimes N$. In particular, if M and N are finitely generated, so is $M \otimes N$.

ii) The notation $x \otimes y$ is inherently ambiguous unless we specify the tensor product to which it belongs. Let M', N' be submodules of M, N respectively, and let $x \in M'$ and $y \in N'$. Then it can happen that $x \otimes y$ as an element of $M \otimes N$ is zero whilst $x \otimes y$ *as an element of* $M' \otimes N'$ is non-zero. For example, take $A = \mathbf{Z}$, $M = \mathbf{Z}$, $N = \mathbf{Z}/2\mathbf{Z}$, and let M' be the submodule $2\mathbf{Z}$ of $\mathbf{Z}$, whilst $N' = N$. Let x be the non-zero element of N and consider $2 \otimes x$. As an element of $M \otimes N$, it is zero because $2 \otimes x = 1 \otimes 2x = 1 \otimes 0 = 0$. But as an element of $M' \otimes N'$ it is non-zero. See the example after (2.18).

However, there is the following result:

Corollary 2.13. *Let $x_i \in M$, $y_i \in N$ be such that $\sum x_i \otimes y_i = 0$ in $M \otimes N$. Then there exist finitely generated submodules M_0 of M and N_0 of N such that $\sum x_i \otimes y_i = 0$ in $M_0 \otimes N_0$.*

Proof. If $\sum x_i \otimes y_i = 0$ in $M \otimes N$, then in the notation of the proof of (2.11) we have $\sum (x_i, y_i) \in D$, and therefore $\sum (x_i, y_i)$ is a finite sum of generators of D. Let M_0 be the submodule of M generated by the x_i and all the elements of M which occur as first coordinates in these generators of D, and define N_0 similarly. Then $\sum x_i \otimes y_i = 0$ as an element of $M_0 \otimes N_0$. ∎

iii) We shall never again need to use the construction of the tensor product given in (2.12), and the reader may safely forget it if he prefers. What is essential to keep in mind is the defining property of the tensor product.

iv) Instead of starting with bilinear mappings we could have started with multilinear mappings $f: M_1 \times \cdots \times M_r \to P$ defined in the same way (i.e., linear in each variable). Following through the proof of (2.12) we should end up with a "multi-tensor product" $T = M_1 \otimes \cdots \otimes M_r$, generated by all products $x_1 \otimes \cdots \otimes x_r$ ($x_i \in M_i, 1 \leqslant i \leqslant r$). The details may safely be left to the reader; the result corresponding to (2.12) is

***Proposition* 2.12*.** *Let $M_1, \ldots, M_r$ be A-modules. Then there exists a pair (T, g) consisting of an A-module T and an A-multilinear mapping $g: M_1 \times \cdots \times M_r \to T$ with the following property:*

Given any A-module P and any A-multilinear mapping $f: M_1 \times \cdots \times M_r \to T$, there exists a unique A-homomorphism $f': T \to P$ such that $f' \circ g = f$.

Moreover, if (T, g) and (T', g') are two pairs with this property, then there exists a unique isomorphism $j: T \to T'$ such that $j \circ g = g'$. ■

There are various so-called "canonical isomorphisms", some of which we state here:

***Proposition* 2.14.** *Let M, N, P be A-modules. Then there exist unique isomorphisms*

i) $M \otimes N \to N \otimes M$

ii) $(M \otimes N) \otimes P \to M \otimes (N \otimes P) \to M \otimes N \otimes P$

iii) $(M \oplus N) \otimes P \to (M \otimes P) \oplus (N \otimes P)$

iv) $A \otimes M \to M$

such that, respectively,

a) $x \otimes y \mapsto y \otimes x$

b) $(x \otimes y) \otimes z \mapsto x \otimes (y \otimes z) \mapsto x \otimes y \otimes z$

c) $(x, y) \otimes z \mapsto (x \otimes z, y \otimes z)$

d) $a \otimes x \mapsto ax$.

Proof. In each case the point is to show that the mappings so described are well defined. The technique is to construct suitable bilinear or multilinear mappings, and use the defining property (2.12) or (2.12*) to infer the existence of homomorphisms of tensor products. We shall prove half of ii) as an example of the method, and leave the rest to the reader.

We shall construct homomorphisms

$$(M \otimes N) \otimes P \xrightarrow{f} M \otimes N \otimes P \xrightarrow{g} (M \otimes N) \otimes P$$

such that $f((x \otimes y) \otimes z) = x \otimes y \otimes z$ and $g(x \otimes y \otimes z) = (x \otimes y) \otimes z$ for all $x \in M, y \in N, z \in P$.

To construct f, fix the element $z \in P$. The mapping $(x, y) \mapsto x \otimes y \otimes z$ ($x \in M, y \in N$) is bilinear in x and y and therefore induces a homomorphism

$f_z: M \otimes N \to M \otimes N \otimes P$ such that $f_z(x \otimes y) = x \otimes y \otimes z$. Next, consider the mapping $(t, z) \mapsto f_z(t)$ of $(M \otimes N) \times P$ into $M \otimes N \otimes P$. This is bilinear in t and z and therefore induces a homomorphism

$$f: (M \otimes N) \otimes P \to M \otimes N \otimes P$$

such that $f((x \otimes y) \otimes z) = x \otimes y \otimes z$.

To construct g, consider the mapping $(x, y, z) \mapsto (x \otimes y) \otimes z$ of $M \times N \times P$ into $(M \otimes N) \otimes P$. This is linear in each variable and therefore induces a homomorphism

$$g: M \otimes N \otimes P \to (M \otimes N) \otimes P$$

such that $g(x \otimes y \otimes z) = (x \otimes y) \otimes z$.

Clearly $f \circ g$ and $g \circ f$ are identity maps, hence f and g are isomorphisms. ∎

Exercise 2.15. Let A, B be rings, let M be an A-module, P a B-module and N an (A, B)-bimodule (that is, N is simultaneously an A-module and a B-module and the two structures are compatible in the sense that $a(xb) = (ax)b$ for all $a \in A$, $b \in B$, $x \in N$). Then $M \otimes_A N$ is naturally a B-module, $N \otimes_B P$ an A-module, and we have

$$(M \otimes_A N) \otimes_B P \cong M \otimes_A (N \otimes_B P).$$

Let $f: M \to M'$, $g: N \to N'$ be homomorphisms of A-modules. Define $h: M \times N \to M' \otimes N'$ by $h(x, y) = f(x) \otimes g(y)$. It is easily checked that h is A-bilinear and therefore induces an A-module homomorphism

$$f \otimes g: M \otimes N \to M' \otimes N'$$

such that

$$(f \otimes g)(x \otimes y) = f(x) \otimes g(y) \qquad (x \in M, \quad y \in N).$$

Let $f': M' \to M''$ and $g': N' \to N''$ be homomorphisms of A-modules. Then clearly the homomorphisms $(f' \circ f) \otimes (g' \circ g)$ and $(f' \otimes g') \circ (f \otimes g)$ agree on all elements of the form $x \otimes y$ in $M \otimes N$. Since these elements generate $M \otimes N$, it follows that

$$(f' \circ f) \otimes (g' \circ g) = (f' \otimes g') \circ (f \otimes g).$$

RESTRICTION AND EXTENSION OF SCALARS

Let $f: A \to B$ be a homomorphism of rings and let N be a B-module. Then N has an A-module structure defined as follows: if $a \in A$ and $x \in N$, then ax is defined to be $f(a)x$. This A-module is said to be obtained from N by *restriction of scalars*. In particular, f defines in this way an A-module structure on B.

Proposition 2.16. *Suppose N is finitely generated as a B-module and that B is finitely generated as an A-module. Then N is finitely generated as an A-module.*

Proof. Let $y_1, \dots, y_n$ generate N over B, and let $x_1, \dots, x_m$ generate B as an A-module. Then the mn products $x_i y_j$ generate N over A. ∎

Now let M be an A-module. Since, as we have just seen, B can be regarded as an A-module, we can form the A-module $M_B = B \otimes_A M$. In fact M_B carries a B-module structure such that $b(b' \otimes x) = bb' \otimes x$ for all $b, b' \in B$ and all $x \in M$. The B-module M_B is said to be obtained from M by *extension of scalars.*

Proposition 2.17. *If M is finitely generated as an A-module, then M_B is finitely generated as a B-module.*

Proof. If $x_1, \dots, x_m$ generate M over A, then the $1 \otimes x_i$ generate M_B over B. ∎

EXACTNESS PROPERTIES OF THE TENSOR PRODUCT

Let $f: M \times N \to P$ be an A-bilinear mapping. For each $x \in M$ the mapping $y \mapsto f(x, y)$ of N into P is A-linear, hence f gives rise to a mapping $M \to \text{Hom}(N, P)$ which is A-linear because f is linear in the variable x. Conversely any A-homomorphism $\phi: M \to \text{Hom}_A(N, P)$ defines a bilinear map, namely $(x, y) \mapsto \phi(x)(y)$. Hence the set S of A-bilinear mappings $M \times N \to P$ is in natural one-to-one correspondence with $\text{Hom}(M, \text{Hom}(N, P))$. On the other hand S is in one-to-one correspondence with $\text{Hom}(M \otimes N, P)$, by the defining property of the tensor product. Hence we have a canonical isomorphism

$$\text{Hom}(M \otimes N, P) \cong \text{Hom}(M, \text{Hom}(N, P)). \tag{1}$$

Proposition 2.18. *Let*

$$M' \xrightarrow{f} M \xrightarrow{g} M'' \to 0 \tag{2}$$

be an exact sequence of A-modules and homomorphisms, and let N be any A-module. Then the sequence

$$M' \otimes N \xrightarrow{f \otimes 1} M \otimes N \xrightarrow{g \otimes 1} M'' \otimes N \to 0 \tag{3}$$

(where 1 denotes the identity mapping on N) is exact.

Proof. Let E denote the sequence (2), and let $E \otimes N$ denote the sequence (3). Let P be any A-module. Since (2) is exact, the sequence $\text{Hom}(E, \text{Hom}(N, P))$ is exact by (2.9); hence by (1) the sequence $\text{Hom}(E \otimes N, P)$ is exact. By (2.9) again, it follows that $E \otimes N$ is exact. ∎

Remarks. i) Let $T(M) = M \otimes N$ and let $U(P) = \text{Hom}(N, P)$. Then (1) takes the form $\text{Hom}(T(M), P) = \text{Hom}(M, U(P))$ for all A-modules M and P. In the language of abstract nonsense, the functor T is the left adjoint of U, and U is the right adjoint of T. The proof of (2.18) shows that any functor which is a left adjoint is right exaet. Likewise any functor which is a right adjoint is left exact.

ii) It is *not* in general true that, if $M' \to M \to M''$ is an exact sequence of A-modules and homomorphisms, the sequence $M' \otimes N \to M \otimes N \to M'' \otimes N$ obtained by tensoring with an arbitrary A-module N is exact.

Example. Take $A = \mathbf{Z}$ and consider the exact sequence $0 \to \mathbf{Z} \xrightarrow{f} \mathbf{Z}$, where $f(x) = 2x$ for all $x \in \mathbf{Z}$. If we tensor with $N = \mathbf{Z}/2\mathbf{Z}$, the sequence $0 \to \mathbf{Z} \otimes N \xrightarrow{f \otimes 1} \mathbf{Z} \otimes N$ is *not* exact, because for any $x \otimes y \in \mathbf{Z} \otimes N$ we have

$$(f \otimes 1)(x \otimes y) = 2x \otimes y = x \otimes 2y = x \otimes 0 = 0,$$

so that $f \otimes 1$ is the zero mapping, whereas $\mathbf{Z} \otimes N \neq 0$.

The functor $T_N : M \mapsto M \otimes_A N$ on the category of A-modules and homomorphisms is therefore not in general exact. If T_N is exact, that is to say if tensoring with N transforms all exact sequences into exact sequences, then N is said to be a *flat* A-module.

Proposition 2.19. *The following are equivalent, for an A-module N:*

i) *N is flat.*

ii) *If $0 \to M' \to M \to M'' \to 0$ is any exact sequence of A-modules, the tensored sequence $0 \to M' \otimes N \to M \otimes N \to M'' \otimes N \to 0$ is exact.*

iii) *If $f: M' \to M$ is injective, then $f \otimes 1: M' \otimes N \to M \otimes N$ is injective.*

iv) *If $f: M' \to M$ is injective and M, M' are finitely generated, then $f \otimes 1: M' \otimes N \to M \otimes N$ is injective.*

Proof. i) $\Leftrightarrow$ ii) by splitting up a long exact sequence into short exact sequences.

ii) $\Leftrightarrow$ iii) by (2.18).

iii) $\Rightarrow$ iv): clear.

iv) $\Rightarrow$ iii). Let $f: M' \to M$ be injective and let $u = \sum x_i' \otimes y_i \in \operatorname{Ker}(f \otimes 1)$, so that $\sum f(x_i') \otimes y_i = 0$ in $M \otimes N$. Let M_0' be the submodule of M' generated by the x_i' and let u_0 denote $\sum x_i' \otimes y_i$ as an element of $M_0' \otimes N$. By (2.14) there exists a finitely generated submodule M_0 of M containing $f(M_0')$ and such that $\sum f(x_i') \otimes y_i = 0$ as an element of $M_0 \otimes N$. If $f_0: M_0' \to M_0$ is the restriction of f, this means that $(f_0 \otimes 1)(u_0) = 0$. Since M_0 and M_0' are finitely generated, $f_0 \otimes 1$ is injective and therefore $u_0 = 0$, hence $u = 0$. ∎

Exercise 2.20. *If $f: A \to B$ is a ring homomorphism and M is a flat A-module, then $M_B = B \otimes_A M$ is a flat B-module.* (Use the canonical isomorphisms (2.14), (2.15).)

ALGEBRAS

Let $f: A \to B$ be a ring homomorphism. If $a \in A$ and $b \in B$, define a product

$$ab = f(a)b.$$

This definition of scalar multiplication makes the ring B into an A-module (it is a particular example of restriction of scalars). Thus B has an A-module structure as well as a ring structure, and these two structures are compatible in a sense which the reader will be able to formulate for himself. The ring B, equipped with this A-module structure, is said to be an *A-algebra.* Thus an A-algebra is, by definition, a ring B together with a ring homomorphism $f: A \to B$.

Remarks. i) In particular, if A is a field K (and $B \neq 0$) then f is injective by (1.2) and therefore K can be canonically identified with its image in B. Thus a K-algebra (K a field) is effectively a ring containing K as a subring.

ii) Let A be any ring. Since A has an identity element there is a unique homomorphism of the ring of integers $\mathbf{Z}$ into A, namely $n \mapsto n.1$. Thus every ring is automatically a $\mathbf{Z}$-algebra.

Let $f: A \to B$, $g: A \to C$ be two ring homomorphisms. An *A-algebra homomorphism* $h: B \to C$ is a ring homomorphism which is also an A-module homomorphism. The reader should verify that h is an A-algebra homomorphism if and only if $h \circ f = g$.

A ring homomorphism $f: A \to B$ is *finite*, and B is a *finite* A-algebra, if B is finitely generated as an A-module. The homomorphism f is *of finite type*, and B is a *finitely-generated A-algebra*, if there exists a finite set of elements $x_1, \dots x_n$ in B such that every element of B can be written as a polynomial in $x_1, \dots, x_n$ with coefficients in $f(A)$; or equivalently if there is an A-algebra homomorphism from a polynomial ring $A[t_1, \dots, t_n]$ *onto* B.

A ring A is said to be *finitely generated* if it is finitely generated as a $\mathbf{Z}$-algebra. This means that there exist finitely many elements $x_1, \dots, x_n$ in A such that every element of A can be written as a polynomial in the x_i with rational integer coefficients.

TENSOR PRODUCT OF ALGEBRAS

Let B, C be two A-algebras, $f: A \to B$, $g: A \to C$ the corresponding homomorphisms. Since B and C are A-modules we may form their tensor product $D = B \otimes_A C$, which is an A-module. We shall now define a multiplication on D.

Consider the mapping $B \times C \times B \times C \to D$ defined by

$$(b, c, b', c') \mapsto bb' \otimes cc'.$$

This is A-linear in each factor and therefore, by (2.12*), induces an A-module homomorphism

$$B \otimes C \otimes B \otimes C \to D,$$

hence by (2.14) an A-module homomorphism

$$D \otimes D \to D$$

and this in turn by (2.11) corresponds to an A-bilinear mapping

$$\mu: D \times D \to D$$

which is such that

$$\mu(b \otimes c, b' \otimes c') = bb' \otimes cc'.$$

Of course, we could have written down this formula directly, but without some such argument as we have given there would be no guarantee that μ was well-defined.

We have therefore defined a multiplication on the tensor product $D = B \otimes_A C$: for elements of the form $b \otimes c$ it is given by

$$(b \otimes c)(b' \otimes c') = bb' \otimes cc',$$

and in general by

$$\left(\sum_i (b_i \otimes c_i)\right)\left(\sum_j (b'_j \otimes c'_j)\right) = \sum_{i,j} (b_i b'_j \otimes c_i c'_j).$$

The reader should check that with this multiplication D is a commutative ring, with identity element $1 \otimes 1$. Furthermore, D is an A-algebra: the mapping $a \mapsto f(a) \otimes g(a)$ is a ring homomorphism $A \to D$.

In fact there is a commutative diagram of ring homomorphisms

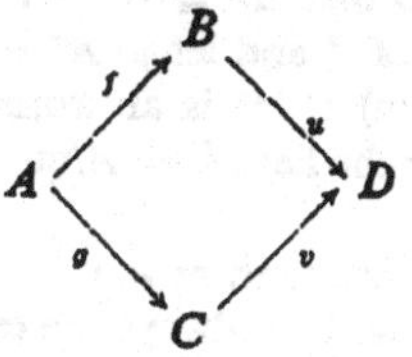

in which u, for example, is defined by $u(b) = b \otimes 1$.

EXERCISES

1. Show that $(\mathbf{Z}/m\mathbf{Z}) \otimes_{\mathbf{Z}} (\mathbf{Z}/n\mathbf{Z}) = 0$ if m, n are coprime.

2. Let A be a ring, $\mathfrak{a}$ an ideal, M an A-module. Show that $(A/\mathfrak{a}) \otimes_A M$ is isomorphic to $M/\mathfrak{a}M$.
[Tensor the exact sequence $0 \to \mathfrak{a} \to A \to A/\mathfrak{a} \to 0$ with M.]

3. Let A be a local ring, M and N finitely generated A-modules. Prove that if $M \otimes N = 0$, then $M = 0$ or $N = 0$.
[Let $\mathfrak{m}$ be the maximal ideal, $k = A/\mathfrak{m}$ the residue field. Let $M_k = k \otimes_A M \cong M/\mathfrak{m}M$ by Exercise 2. By Nakayama's lemma, $M_k = 0 \Rightarrow M = 0$. But $M \otimes_A N = 0 \Rightarrow (M \otimes_A N)_k = 0 \Rightarrow M_k \otimes_k N_k = 0 \Rightarrow M_k = 0$ or $N_k = 0$, since M_k, N_k are vector spaces over a field.]

4. Let M_i $(i \in I)$ be any family of A-modules, and let M be their direct sum. Prove that M is flat $\Leftrightarrow$ each M_i is flat.

5. Let $A[x]$ be the ring of polynomials in one indeterminate over a ring A. Prove that $A[x]$ is a flat A-algebra. [Use Exercise 4.]

6. For any A-module, let $M[x]$ denote the set of all polynomials in x with coefficients in M, that is to say expressions of the form

$$m_0 + m_1x + \cdots + m_rx^r \qquad (m_i \in M).$$

Defining the product of an element of $A[x]$ and an element of $M[x]$ in the obvious way, show that $M[x]$ is an $A[x]$-module.

Show that $M[x] \cong A[x] \otimes_A M$.

7. Let $\mathfrak{p}$ be a prime ideal in A. Show that $\mathfrak{p}[x]$ is a prime ideal in $A[x]$. If $\mathfrak{m}$ is a maximal ideal in A, is $\mathfrak{m}[x]$ a maximal ideal in $A[x]$?

8. i) If M and N are flat A-modules, then so is $M \otimes_A N$.
ii) If B is a flat A-algebra and N is a flat B-module, then N is flat as an A-module.

9. Let $0 \to M' \to M \to M'' \to 0$ be an exact sequence of A-modules. If M' and M'' are finitely generated, then so is M.

10. Let A be a ring, $\mathfrak{a}$ an ideal contained in the Jacobson radical of A; let M be an A-module and N a finitely generated A-module, and let $u: M \to N$ be a homomorphism. If the induced homomorphism $M/\mathfrak{a}M \to N/\mathfrak{a}N$ is surjective, then u is surjective.

11. Let A be a ring $\neq 0$. Show that $A^m \cong A^n \Rightarrow m = n$.
[Let $\mathfrak{m}$ be a maximal ideal of A and let $\phi: A^m \to A^n$ be an isomorphism. Then $1 \otimes \phi: (A/\mathfrak{m}) \otimes A^m \to (A/\mathfrak{m}) \otimes A^n$ is an isomorphism between vector spaces of dimensions m and n over the field $k = A/\mathfrak{m}$. Hence $m = n$.] (Cf. Chapter 3, Exercise 15.)

If $\phi: A^m \to A^n$ is surjective, then $m \geq n$.

If $\phi: A^m \to A^n$ is injective, is it always the case that $m \leq n$?

12. Let M be a finitely generated A-module and $\phi: M \to A^n$ a surjective homomorphism. Show that Ker (ϕ) is finitely generated.
[Let $e_1, \ldots, e_n$ be a basis of A^n and choose $u_i \in M$ such that $\phi(u_i) = e_i$ $(1 \leq i \leq n)$. Show that M is the direct sum of Ker (ϕ) and the submodule generated by $u_1, \ldots, u_n$.]

13. Let $f: A \to B$ be a ring homomorphism, and let N be a B-module. Regarding N as an A-module by restriction of scalars, form the B-module $N_B = B \otimes_A N$. Show that the homomorphism $g: N \to N_B$ which maps y to $1 \otimes y$ is injective and that $g(N)$ is a direct summand of N_B.
[Define $p: N_B \to N$ by $p(b \otimes y) = by$, and show that $N_B = \text{Im}(g) \oplus \text{Ker}(p)$.]

Direct limits

14. A partially ordered set I is said to be a *directed* set if for each pair i, j in I there exists $k \in I$ such that $i \leq k$ and $j \leq k$.

Let A be a ring, let I be a directed set and let $(M_i)_{i \in I}$ be a family of A-modules indexed by I. For each pair i, j in I such that $i \leq j$, let $\mu_{ij}: M_i \to M_j$ be an A-homomorphism, and suppose that the following axioms are satisfied:

(1) μ_{ii} is the identity mapping of M_i, for all $i \in I$;

(2) $\mu_{ik} = \mu_{jk} \circ \mu_{ij}$ whenever $i \leqslant j \leqslant k$.

Then the modules M_i and homomorphisms μ_{ij} are said to form a *direct system* $\mathbf{M} = (M_i, \mu_{ij})$ over the directed set I.

We shall construct an A-module M called the *direct limit* of the direct system $\mathbf{M}$. Let C be the direct sum of the M_i, and identify each module M_i with its canonical image in C. Let D be the submodule of C generated by all elements of the form $x_i - \mu_{ij}(x_i)$ where $i \leqslant j$ and $x_i \in M_i$. Let $M = C/D$, let $\mu: C \to M$ be the projection and let μ_i be the restriction of μ to M_i.

The module M, or more correctly the pair consisting of M and the family of homomorphisms $\mu_i: M_i \to M$, is called the *direct limit* of the direct system $\mathbf{M}$, and is written $\varinjlim M_i$. From the construction it is clear that $\mu_i = \mu_j \circ \mu_{ij}$ whenever $i \leqslant j$.

15. In the situation of Exercise 14, show that every element of M can be written in the form $\mu_i(x_i)$ for some $i \in I$ and some $x_i \in M_i$.

Show also that if $\mu_i(x_i) = 0$ then there exists $j \geqslant i$ such that $\mu_{ij}(x_i) = 0$ in M_j.

16. Show that the direct limit is characterized (up to isomorphism) by the following property. Let N be an A-module and for each $i \in I$ let $\alpha_i: M_i \to N$ be an A-module homomorphism such that $\alpha_i = \alpha_j \circ \mu_{ij}$ whenever $i \leqslant j$. Then there exists a unique homomorphism $\alpha: M \to N$ such that $\alpha_i = \alpha \circ \mu_i$ for all $i \in I$.

17. Let $(M_i)_{i \in I}$ be a family of submodules of an A-module, such that for each pair of indices i, j in I there exists $k \in I$ such that $M_i + M_j \subseteq M_k$. Define $i \leqslant j$ to mean $M_i \subseteq M_j$ and let $\mu_{ij}: M_i \to M_j$ be the embedding of M_i in M_j. Show that

$$\varinjlim M_i = \sum M_i = \bigcup M_i.$$

In particular, any A-module is the direct limit of its finitely generated submodules.

18. Let $\mathbf{M} = (M_i, \mu_{ij})$, $\mathbf{N} = (N_i, \nu_{ij})$ be direct systems of A-modules over the same directed set. Let M, N be the direct limits and $\mu_i: M_i \to M$, $\nu_i: N_i \to N$ the associated homomorphisms.

A *homomorphism* $\mathbf{\Phi}: \mathbf{M} \to \mathbf{N}$ is by definition a family of A-module homomorphisms $\phi_i: M_i \to N_i$ such that $\phi_j \circ \mu_{ij} = \nu_{ij} \circ \phi_i$ whenever $i \leqslant j$. Show that $\mathbf{\Phi}$ defines a unique homomorphism $\phi = \varinjlim \phi_i: M \to N$ such that $\phi \circ \mu_i = \nu_i \circ \phi_i$ for all $i \in I$.

19. A sequence of direct systems and homomorphisms

$$\mathbf{M} \to \mathbf{N} \to \mathbf{P}$$

is *exact* if the corresponding sequence of modules and module homomorphisms is exact for each $i \in I$. Show that the sequence $M \quad N \to P$ of direct limits is then exact. [Use Exercise 15.]

Tensor products commute with direct limits

20. Keeping the same notation as in Exercise 14, let N be any A-module. Then $(M_i \otimes N, \mu_{ij} \otimes 1)$ is a direct system; let $P = \varinjlim (M_i \otimes N)$ be its direct limit.

For each $i \in I$ we have a homomorphism $\mu_i \otimes 1: M_i \otimes N \to M \otimes N$, hence by Exercise 16 a homomorphism $\psi: P \to M \otimes N$. Show that ψ is an isomorphism, so that

$$\varinjlim (M_i \otimes N) \cong (\varinjlim M_i) \otimes N.$$

[For each $i \in I$, let $g_i: M_i \times N \to M_i \otimes N$ be the canonical bilinear mapping. Passing to the limit we obtain a mapping $g: M \times N \to P$. Show that g is A-bilinear and hence define a homomorphism $\phi: M \otimes N \to P$. Verify that $\phi \circ \psi$ and $\psi \circ \phi$ are identity mappings.]

21. Let $(A_i)_{i \in I}$ be a family of rings indexed by a directed set I, and for each pair $i \leqslant j$ in I let $\alpha_{ij}: A_i \to A_j$ be a ring homomorphism, satisfying conditions (1) and (2) of Exercise 14. Regarding each A_i as a $\mathbf{Z}$-module we can then form the direct limit $A = \varinjlim A_i$. Show that A inherits a ring structure from the A_i so that the mappings $A_i \to A$ are ring homomorphisms. The ring A is the *direct limit* of the system (A_i, α_{ij}).

    If $A = 0$ prove that $A_i = 0$ for some $i \in I$. [Remember that all rings have identity elements!]

22. Let (A_i, α_{ij}) be a direct system of rings and let $\mathfrak{N}_i$ be the nilradical of A_i. Show that $\varinjlim \mathfrak{N}_i$ is the nilradical of $\varinjlim A_i$.

    If each A_i is an integral domain, then $\varinjlim A_i$ is an integral domain.

23. Let $(B_\lambda)_{\lambda \in \Lambda}$ be a family of A-algebras. For each finite subset of Λ let B_J denote the tensor product (over A) of the B_λ for $\lambda \in J$. If J' is another finite subset of Λ and $J \subseteq J'$, there is a canonical A-algebra homomorphism $B_J \to B_{J'}$. Let B denote the direct limit of the rings B_J as J runs through all finite subsets of Λ. The ring B has a natural A-algebra structure for which the homomorphisms $B_J \to B$ are A-algebra homomorphisms. The A-algebra B is the *tensor product* of the family $(B_\lambda)_{\lambda \in \Lambda}$.

Flatness and Tor

In these Exercises it will be assumed that the reader is familiar with the definition and basic properties of the Tor functor.

24. If M is an A-module, the following are equivalent:

    i) M is flat;

    ii) $\mathrm{Tor}_n^A(M, N) = 0$ for all $n > 0$ and all A-modules N;

    iii) $\mathrm{Tor}_1^A(M, N) = 0$ for all A-modules N.

    [To show that (i) $\Rightarrow$ (ii), take a free resolution of N and tensor it with M. Since M is flat, the resulting sequence is exact and therefore its homology groups, which are the $\mathrm{Tor}_n^A(M, N)$, are zero for $n > 0$. To show that (iii) $\Rightarrow$ (i), let $0 \to N' \to N \to N'' \to 0$ be an exact sequence. Then, from the Tor exact sequence,

    $$\mathrm{Tor}_1(M, N'') \to M \otimes N' \to M \otimes N \to M \otimes N'' \to 0$$

    is exact. Since $\mathrm{Tor}_1(M, N'') = 0$ it follows that M is flat.]

25. Let $0 \to N' \to N \to N'' \to 0$ be an exact sequence, with N'' flat. Then N' is flat $\Leftrightarrow$ N is flat. [Use Exercise 24 and the Tor exact sequence.]

26. Let N be an A-module. Then N is flat $\Leftrightarrow$ $\mathrm{Tor}_1(A/\mathfrak{a}, N) = 0$ for all finitely generated ideals $\mathfrak{a}$ in A.

[Show first that N is flat if $\mathrm{Tor}_1(M, N) = 0$ for all *finitely generated* A-modules M, by using (2.19). If M is finitely generated, let $x_1, \ldots, x_n$ be a set of generators of M, and let M_i be the submodule generated by $x_1, \ldots, x_i$. By considering the successive quotients M_i/M_{i-1} and using Exercise 25, deduce that N is flat if $\mathrm{Tor}_1(M, N) = 0$ for all *cyclic* A-modules M, i.e., all M generated by a single element, and therefore of the form $A/\mathfrak{a}$ for some ideal $\mathfrak{a}$. Finally use (2.19) again to reduce to the case where $\mathfrak{a}$ is a finitely generated ideal.]

27. A ring A is *absolutely flat* if every A-module is flat. Prove that the following are equivalent:

i) A is absolutely flat.

ii) Every principal ideal is idempotent.

iii) Every finitely generated ideal is a direct summand of A.

[i) $\Rightarrow$ ii). Let $x \in A$. Then $A/(x)$ is a flat A-module, hence in the diagram

$$\begin{array}{ccc} (x) \otimes A & \xrightarrow{\beta} & (x) \otimes A/(x) \\ \downarrow & & \downarrow{\scriptstyle \alpha} \\ A & \rightarrow & A/(x) \end{array}$$

the mapping α is injective. Hence $\mathrm{Im}(\beta) = 0$, hence $(x) = (x^2)$. ii) $\Rightarrow$ iii). Let $x \in A$. Then $x = ax^2$ for some $a \in A$, hence $e = ax$ is idempotent and we have $(e) = (x)$. Now if e, f are idempotents, then $(e, f) = (e + f - ef)$. Hence every finitely generated ideal is principal, and generated by an idempotent e, hence is a direct summand because $A = (e) \oplus (1 - e)$. iii) $\Rightarrow$ i). Use the criterion of Exercise 26.]

28. A Boolean ring is absolutely flat. The ring of Chapter 1, Exercise 7 is absolutely flat. Every homomorphic image of an absolutely flat ring is absolutely flat. If a local ring is absolutely flat, then it is a field.

If A is absolutely flat, every non-unit in A is a zero-divisor.

3

Rings and Modules of Fractions

The formation of rings of fractions and the associated process of localization are perhaps the most important technical tools in commutative algebra. They correspond in the algebro-geometric picture to concentrating attention on an open set or near a point, and the importance of these notions should be self-evident. This chapter gives the definitions and simple properties of the formation of fractions.

The procedure by which one constructs the rational field **Q** from the ring of integers **Z** (and embeds **Z** in **Q**) extends easily to any integral domain A and produces the *field of fractions* of A. The construction consists in taking all ordered pairs (a, s) where $a, s \in A$ and $s \neq 0$, and setting up an equivalence relation between such pairs:

$$(a, s) \equiv (b, t) \Leftrightarrow at - bs = 0.$$

This works only if A is an integral domain, because the verification that the relation is transitive involves canceling, i.e. the fact that A has no zero-divisor $\neq 0$. However, it can be generalized as follows:

Let A be any ring. A *multiplicatively closed subset* of A is a subset S of A such that $1 \in S$ and S is closed under multiplication: in other words S is a sub-semigroup of the multiplicative semigroup of A. Define a relation $\equiv$ on $A \times S$ as follows:

$$(a, s) \equiv (b, t) \Leftrightarrow (at - bs)u = 0 \text{ for some } u \in S.$$

Clearly this relation is reflexive and symmetric. To show that it is transitive, suppose $(a, s) \equiv (b, t)$ and $(b, t) \equiv (c, u)$. Then there exist v, w in S such that $(at - bs)v = 0$ and $(bu - ct)w = 0$. Eliminate b from these two equations and we have $(au - cs)tvw = 0$. Since S is closed under multiplication, we have $tvw \in S$, hence $(a, s) \equiv (c, u)$. Thus we have an equivalence relation. Let a/s denote the equivalence class of (a, s), and let $S^{-1}A$ denote the set of equivalence classes. We put a ring structure on $S^{-1}A$ by defining addition and multiplication of these "fractions" a/s in the same way as in elementary algebra: that is,

$$(a/s) + (b/t) = (at + bs)/st,$$
$$(a/s)(b/t) = ab/st.$$

Exercise. *Verify that these definitions are independent of the choices of representatives (a, s) and (b, t), and that $S^{-1}A$ satisfies the axioms of a commutative ring with identity.*

We also have a ring homomorphism $f: A \to S^{-1}A$ defined by $f(x) = x/1$. This is *not* in general injective.

Remark. If A is an integral domain and $S = A - \{0\}$, then $S^{-1}A$ is the field of fractions of A.

The ring $S^{-1}A$ is called the *ring of fractions* of A with respect to S. It has a *universal property*:

Proposition 3.1. *Let $g: A \to B$ be a ring homomorphism such that $g(s)$ is a unit in B for all $s \in S$. Then there exists a unique ring homomorphism $h: S^{-1}A \to B$ such that $g = h \circ f$.*

Proof. i) *Uniqueness.* If h satisfies the conditions, then $h(a/1) = hf(a) = g(a)$ for all $a \in A$; hence, if $s \in S$,

$$h(1/s) = h((s/1)^{-1}) = h(s/1)^{-1} = g(s)^{-1}$$

and therefore $h(a/s) = h(a/1) \cdot h(1/s) = g(a)g(s)^{-1}$, so that h is uniquely determined by g.

ii) *Existence.* Let $h(a/s) = g(a)g(s)^{-1}$. Then h will clearly be a ring homomorphism provided that it is well-defined. Suppose then that $a/s = a'/s'$; then there exists $t \in S$ such that $(as' - a's)t = 0$, hence

$$(g(a)g(s') - g(a')g(s))g(t) = 0;$$

now $g(t)$ is a unit in B, hence $g(a)g(s)^{-1} = g(a')g(s')^{-1}$. ∎

The ring $S^{-1}A$ and the homomorphism $f: A \to S^{-1}A$ have the following properties:

1) $s \in S \Rightarrow f(s)$ is a unit in $S^{-1}A$;
2) $f(a) = 0 \Rightarrow as = 0$ for some $s \in S$;
3) Every element of $S^{-1}A$ is of the form $f(a)f(s)^{-1}$ for some $a \in A$ and some $s \in S$.

Conversely, these three conditions determine the ring $S^{-1}A$ up to isomorphism. Precisely:

Corollary 3.2. *If $g: A \to B$ is a ring homomorphism such that*

i) $s \in S \Rightarrow g(s)$ *is a unit in* B;

ii) $g(a) = 0 \Rightarrow as = 0$ *for some* $s \in S$;

iii) *Every element of B is of the form $g(a)g(s)^{-1}$; then there is a unique isomorphism $h: S^{-1}A \to B$ such that $g = h \circ f$.*

Proof. By (3.1) we have to show that $h: S^{-1}A \to B$, defined by

$$h(a/s) = g(a)g(s)^{-1}$$

(this definition uses i)) is an isomorphism. By iii), h is surjective. To show h is injective, look at the kernel of h: if $h(a/s) = 0$, then $g(a) = 0$, hence by ii) we have $at = 0$ for some $t \in S$, hence $(a, s) \equiv (0, 1)$, i.e., $a/s = 0$ in $S^{-1}A$. ∎

Examples. 1) Let $\mathfrak{p}$ be a prime ideal of A. Then $S = A - \mathfrak{p}$ is multiplicatively closed (in fact $A - \mathfrak{p}$ is multiplicatively closed $\Leftrightarrow \mathfrak{p}$ is prime). We write $A_{\mathfrak{p}}$ for $S^{-1}A$ in this case. The elements a/s with $a \in \mathfrak{p}$ form an ideal $\mathfrak{m}$ in $A_{\mathfrak{p}}$. If $b/t \notin \mathfrak{m}$, then $b \notin \mathfrak{p}$, hence $b \in S$ and therefore b/t is a unit in $A_{\mathfrak{p}}$. It follows that if $\mathfrak{a}$ is an ideal in $A_{\mathfrak{p}}$ and $\mathfrak{a} \nsubseteq \mathfrak{m}$, then $\mathfrak{a}$ contains a unit and is therefore the whole ring. Hence $\mathfrak{m}$ is the only maximal ideal in $A_{\mathfrak{p}}$; in other words, $A_{\mathfrak{p}}$ is a *local ring*.

The process of passing from A to $A_{\mathfrak{p}}$ is called *localization* at $\mathfrak{p}$.

2) $S^{-1}A$ is the zero ring $\Leftrightarrow 0 \in S$.

3) Let $f \in A$ and let $S = \{f^n\}_{n \geq 0}$. We write A_f for $S^{-1}A$ in this case.

4) Let $\mathfrak{a}$ be any ideal in A, and let $S = 1 + \mathfrak{a} =$ set of all $1 + x$ where $x \in \mathfrak{a}$. Clearly S is multiplicatively closed.

5) Special cases of 1) and 3):

i) $A = \mathbf{Z}$, $\mathfrak{p} = (p)$, p a prime number; $A_{\mathfrak{p}} =$ set of all rational numbers m/n where n is prime to p; if $f \in \mathbf{Z}$ and $f \neq 0$, then A_f is the set of all rational numbers whose denominator is a power of f.

ii) $A = k[t_1, \ldots, t_n]$, where k is a field and the t_i are independent indeterminates, $\mathfrak{p}$ a prime ideal in A. Then $A_{\mathfrak{p}}$ is the ring of all rational functions f/g, where $g \notin \mathfrak{p}$. If V is the variety defined by the ideal $\mathfrak{p}$, that is to say the set of all $x = (x_1, \ldots, x_n) \in k^n$ such that $f(x) = 0$ whenever $f \in \mathfrak{p}$, then (provided k is infinite) $A_{\mathfrak{p}}$ can be identified with the ring of all rational functions on k^n which are defined at almost all points of V; it is the local ring of k^n *along the variety* V. This is the prototype of the local rings which arise in algebraic geometry.

The construction of $S^{-1}A$ can be carried through with an A-module M in place of the ring A. Define a relation $\equiv$ on $M \times S$ as follows:

$$(m, s) \equiv (m', s') \Leftrightarrow \exists t \in S \text{ such that } t(sm' - s'm) = 0.$$

As before, this is an equivalence relation. Let m/s denote the equivalence class of the pair (m, s), let $S^{-1}M$ denote the set of such fractions, and make $S^{-1}M$ into an $S^{-1}A$-module with the obvious definitions of addition and scalar multiplication. As in Examples 1) and 3) above, we write $M_{\mathfrak{p}}$ instead of $S^{-1}M$ when $S = A - \mathfrak{p}$ ($\mathfrak{p}$ prime) and M_f when $S = \{f^n\}_{n \geq 0}$.

Let $u: M \to N$ be an A-module homomorphism. Then it gives rise to an $S^{-1}A$-module homomorphism $S^{-1}u: S^{-1}M \to S^{-1}N$, namely $S^{-1}u$ maps m/s to $u(m)/s$. We have $S^{-1}(v \circ u) = (S^{-1}v) \circ (S^{-1}u)$.

Proposition 3.3. *The operation S^{-1} is exact, i.e., if $M' \xrightarrow{f} M \xrightarrow{g} M''$ is exact at M, then $S^{-1}M' \xrightarrow{S^{-1}f} S^{-1}M \xrightarrow{S^{-1}g} S^{-1}M''$ is exact at $S^{-1}M$.*

Proof. We have $g \circ f = 0$, hence $S^{-1}g \circ S^{-1}f = S^{-1}(0) = 0$, hence $\text{Im}\,(S^{-1}f) \subseteq \text{Ker}\,(S^{-1}g)$. To prove the reverse inclusion, let $m/s \in \text{Ker}\,(S^{-1}g)$, then $g(m)/s = 0$ in $S^{-1}M''$, hence there exists $t \in S$ such that $tg(m) = 0$ in M''. But $tg(m) = g(tm)$ since g is an A-module homomorphism, hence $tm \in \text{Ker}\,(g) = \text{Im}\,(f)$ and therefore $tm = f(m')$ for some $m' \in M'$. Hence in $S^{-1}M$ we have $m/s = f(m')/st = (S^{-1}f)(m'/st) \in \text{Im}\,(S^{-1}f)$. Hence $\text{Ker}\,(S^{-1}g) \subseteq \text{Im}\,(S^{-1}f)$. ∎

In particular, it follows from (3.3) that if M' is a submodule of M, the mapping $S^{-1}M' \to S^{-1}M$ is *injective* and therefore $S^{-1}M'$ can be regarded as a submodule of $S^{-1}M$. With this convention,

Corollary 3.4. *Formation of fractions commutes with formation of finite sums, finite intersections and quotients. Precisely, if N, P are submodules of an A-module M, then*

i) $S^{-1}(N + P) = S^{-1}(N) + S^{-1}(P)$

ii) $S^{-1}(N \cap P) = S^{-1}(N) \cap S^{-1}(P)$

iii) *the $S^{-1}A$-modules $S^{-1}(M/N)$ and $(S^{-1}M)/(S^{-1}N)$ are isomorphic.*

Proof. i) follows readily from the definitions and ii) is easy to verify: if $y/s = z/t$ $(y \in N, z \in P, s, t \in S)$ then $u(ty - sz) = 0$ for some $u \in S$, hence $w = uty = usz \in N \cap P$ and therefore $y/s = w/stu \in S^{-1}(N \cap P)$. Consequently $S^{-1}N \cap S^{-1}P \subseteq S^{-1}(N \cap P)$, and the reverse inclusion is obvious.

iii) Apply S^{-1} to the exact sequence $0 \to N \to M \to M/N \to 0$. ∎

Proposition 3.5. *Let M be an A-module. Then the $S^{-1}A$ modules $S^{-1}M$ and $S^{-1}A \otimes_A M$ are isomorphic; more precisely, there exists a unique isomorphism $f\colon S^{-1}A \otimes_A M \to S^{-1}M$ for which*

$$f((a/s) \otimes m) = am/s \text{ for all } a \in A, m \in M, s \in S. \tag{1}$$

Proof. The mapping $S^{-1}A \times M \to S^{-1}M$ defined by

$$(a/s, m) \mapsto am/s$$

is A-bilinear, and therefore by the universal property (2.12) of the tensor product induces an A-homomorphism

$$f\colon S^{-1}A \otimes_A M \to S^{-1}M$$

satisfying (1). Clearly f is surjective, and is uniquely defined by (1).

Let $\sum_i (a_i/s_i) \otimes m_i$ be any element of $S^{-1}A \otimes M$. If $s = \prod_i s_i \in S$, $t_i = \prod_{j \neq i} s_j$, we have

$$\sum_i \frac{a_i}{s_i} \otimes m_i = \sum_i \frac{a_i t_i}{s} \otimes m = \sum_i \frac{1}{s} \otimes a_i t_i m = \frac{1}{s} \otimes \sum_i a_i t_i m,$$

so that every element of $S^{-1}A \otimes M$ is of the form $(1/s) \otimes m$. Suppose that $f((1/s) \otimes m) = 0$. Then $m/s = 0$, hence $tm = 0$ for some $t \in S$, and therefore

$$\frac{1}{s} \otimes m = \frac{t}{st} \otimes m = \frac{1}{st} \otimes tm = \frac{1}{st} \otimes 0 = 0.$$

Hence f is injective and therefore an isomorphism. ■

Corollary 3.6. $S^{-1}A$ is a flat A-module.

Proof. (3.3), (3.5). ■

Proposition 3.7. If M, N are A-modules, there is a unique isomorphism of $S^{-1}A$-modules $f: S^{-1}M \otimes_{S^{-1}A} S^{-1}N \to S^{-1}(M \otimes_A N)$ such that

$$f((m/s) \otimes (n/t)) = (m \otimes n)/st.$$

In particular, if $\mathfrak{p}$ is any prime ideal, then

$$M_\mathfrak{p} \otimes_{A_\mathfrak{p}} N_\mathfrak{p} \cong (M \otimes_A N)_\mathfrak{p}$$

as $A_\mathfrak{p}$-modules.

Proof. Use (3.5) and the canonical isomorphisms of Chapter 2. ■

LOCAL PROPERTIES

A property P of a ring A (or of an A-module M) is said to be a *local property* if the following is true:

A (or M) has $P \Leftrightarrow A_\mathfrak{p}$ (or $M_\mathfrak{p}$) has P, for each prime ideal $\mathfrak{p}$ of A. The following propositions give examples of local properties:

Proposition 3.8. Let M be an A-module. Then the following are equivalent:

i) $M = 0$;

ii) $M_\mathfrak{p} = 0$ *for all prime ideals* $\mathfrak{p}$ *of* A;

iii) $M_\mathfrak{m} = 0$ *for all maximal ideals* $\mathfrak{m}$ *of* A.

Proof. Clearly i) ⇒ ii) ⇒ iii). Suppose iii) satisfied and $M \neq 0$. Let x be a non-zero element of M, and let $\mathfrak{a} = \text{Ann}(x)$; $\mathfrak{a}$ is an ideal $\neq (1)$, hence is contained in a maximal ideal $\mathfrak{m}$ by (1.4). Consider $x/1 \in M_\mathfrak{m}$. Since $M_\mathfrak{m} = 0$ we have $x/1 = 0$, hence x is killed by some element of $A - \mathfrak{m}$; but this is impossible since $\text{Ann}(x) \subseteq \mathfrak{m}$. ■

Proposition 3.9. Let $\phi: M \to N$ be an A-module homomorphism. Then the following are equivalent:

i) ϕ *is injective;*

ii) $\phi_\mathfrak{p}: M_\mathfrak{p} \to N_\mathfrak{p}$ *is injective for each prime ideal* $\mathfrak{p}$;

iii) $\phi_\mathfrak{m}: M_\mathfrak{m} \to N_\mathfrak{m}$ *is injective for each maximal ideal* $\mathfrak{m}$.

Similarly with "injective" replaced by "surjective" throughout.

Proof. i) ⇒ ii). $0 \to M \to N$ is exact, hence $0 \to M_\mathfrak{p} \to N_\mathfrak{p}$ is exact, i.e., $\phi_\mathfrak{p}$ is injective.

ii) ⇒ iii) because a maximal ideal is prime.

iii) ⇒ i). Let $M' = \text{Ker}(\phi)$, then the sequence $0 \to M' \to M \to N$ is exact, hence $0 \to M'_\mathfrak{m} \to M_\mathfrak{m} \to N_\mathfrak{m}$ is exact by (3.3) and therefore $M'_\mathfrak{m} \cong \text{Ker}(\phi_\mathfrak{m}) = 0$ since $\phi_\mathfrak{m}$ is injective. Hence $M' = 0$ by (3.8), hence ϕ is injective.

For the other part of the proposition, just reverse all the arrows. ∎

Flatness is a local property:

Proposition 3.10. *For any A-module M, the following statements are equivalent:*

i) *M is a flat A-module:*

ii) *$M_\mathfrak{p}$ is a flat $A_\mathfrak{p}$-module for each prime ideal $\mathfrak{p}$;*

iii) *$M_\mathfrak{m}$ is a flat $A_\mathfrak{m}$-module for each maximal ideal $\mathfrak{m}$.*

Proof. i) ⇒ ii) by (3.5) and (2.20).

ii) ⇒ iii) O.K.

iii) ⇒ i). If $N \to P$ is a homomorphism of A-modules, and $\mathfrak{m}$ is any maximal ideal of A, then

$$\begin{aligned} N \to P \text{ injective} &\Rightarrow N_\mathfrak{m} \to P_\mathfrak{m} \text{ injective, by (3.9)} \\ &\Rightarrow N_\mathfrak{m} \otimes_{A_\mathfrak{m}} M_\mathfrak{m} \to P_\mathfrak{m} \otimes_{A_\mathfrak{m}} M_\mathfrak{m} \text{ injective, by (2.19)} \\ &\Rightarrow (N \otimes_A M)_\mathfrak{m} \to (P \otimes_A M)_\mathfrak{m} \text{ injective, by (3.7)} \\ &\Rightarrow N \otimes_A M \to P \otimes_A M \text{ injective, by (3.9).} \end{aligned}$$

Hence M is flat by (2.19). ∎

EXTENDED AND CONTRACTED IDEALS IN RINGS OF FRACTIONS

Let A be a ring, S a multiplicatively closed subset of A and $f: A \to S^{-1}A$ the natural homomorphism, defined by $f(a) = a/1$. Let C be the set of contracted ideals in A, and let E be the set of extended ideals in $S^{-1}A$ (cf. (1.17)). If $\mathfrak{a}$ is an ideal in A, its extension $\mathfrak{a}^e$ in $S^{-1}A$ is $S^{-1}\mathfrak{a}$ (for any $y \in \mathfrak{a}^e$ is of the form $\sum a_i/s_i$, where $a_i \in \mathfrak{a}$ and $s_i \in S$; bring this fraction to a common denominator).

Proposition 3.11. i) *Every ideal in $S^{-1}A$ is an extended ideal.*

ii) *If $\mathfrak{a}$ is an ideal in A, then $\mathfrak{a}^{ec} = \bigcup_{s \in S} (\mathfrak{a} : s)$. Hence $\mathfrak{a}^e = (1)$ if and only if $\mathfrak{a}$ meets S.*

iii) *$\mathfrak{a} \in C \Leftrightarrow$ no element of S is a zero-divisor in $A/\mathfrak{a}$.*

iv) *The prime ideals of $S^{-1}A$ are in one-to-one correspondence ($\mathfrak{p} \leftrightarrow S^{-1}\mathfrak{p}$) with the prime ideals of A which don't meet S.*

v) *The operation S^{-1} commutes with formation of finite sums, products, intersections and radicals.*

Proof. i) Let $\mathfrak{b}$ be an ideal in $S^{-1}A$, and let $x/s \in \mathfrak{b}$. Then $x/1 \in \mathfrak{b}$, hence $x \in \mathfrak{b}^c$ and therefore $x/s \in \mathfrak{b}^{ce}$. Since $\mathfrak{b} \supseteq \mathfrak{b}^{ce}$ in any case (1.17), it follows that $\mathfrak{b} = \mathfrak{b}^{ce}$.

ii) $x \in \mathfrak{a}^{ec} = (S^{-1}\mathfrak{a})^c \Leftrightarrow x/1 = a/s$ for some $a \in \mathfrak{a}, s \in S \Leftrightarrow (xs - a)t = 0$ for some $t \in S \Leftrightarrow xst \in \mathfrak{a} \Leftrightarrow x \in \bigcup_{s \in S} (\mathfrak{a}:s)$.

iii) $\mathfrak{a} \in C \Leftrightarrow \mathfrak{a}^{ec} \subseteq \mathfrak{a} \Leftrightarrow (sx \in \mathfrak{a}$ for some $s \in S \Rightarrow x \in \mathfrak{a}) \Leftrightarrow$ no $s \in S$ is a zero-divisor in $A/\mathfrak{a}$.

iv) If $\mathfrak{q}$ is a prime ideal in $S^{-1}A$, then $\mathfrak{q}^c$ is a prime ideal in A (this much is true for any ring homomorphism). Conversely, if $\mathfrak{p}$ is a prime ideal in A, then $A/\mathfrak{p}$ is an integral domain; if $\bar{S}$ is the image of S in $A/\mathfrak{p}$, we have $S^{-1}A/S^{-1}\mathfrak{p} \cong \bar{S}^{-1}(A/\mathfrak{p})$ which is either 0 or else is contained in the field of fractions of $A/\mathfrak{p}$ and is therefore an integral domain, and therefore $S^{-1}\mathfrak{p}$ is either prime or is the unit ideal; by i) the latter possibility occurs if and only if $\mathfrak{p}$ meets S.

v) For sums and products, this follows from (1.18); for intersections, from (3.4). As to radicals, we have $S^{-1}r(\mathfrak{a}) \subseteq r(S^{-1}\mathfrak{a})$ from (1.18), and the proof of the reverse inclusion is a routine verification which we leave to the reader. ∎

Remarks. 1) If $\mathfrak{a}, \mathfrak{b}$ are ideals of A, the formula

$$S^{-1}(\mathfrak{a}:\mathfrak{b}) = (S^{-1}\mathfrak{a}:S^{-1}\mathfrak{b})$$

is true provided the ideal $\mathfrak{b}$ is finitely generated: see (3.15).

2) The proof in (1.8) that if $f \in A$ is not nilpotent there is a prime ideal of A which does not contain f can be expressed more concisely in the language of rings of fractions. Since the set $S = (f^n)_{n \geq 0}$ does not contain 0, the ring $S^{-1}A = A_f$ is not the zero ring and therefore by (1.3) has a maximal ideal, whose contraction in A is a prime ideal $\mathfrak{p}$ which does not meet S by (3.11); hence $f \notin \mathfrak{p}$.

Corollary 3.12. *If $\mathfrak{N}$ is the nilradical of A, the nilradical of $S^{-1}A$ is $S^{-1}\mathfrak{N}$.* ∎

Corollary 3.13. *If $\mathfrak{p}$ is a prime ideal of A, the prime ideals of the local ring $A_\mathfrak{p}$ are in one-to-one correspondence with the prime ideals of A contained in $\mathfrak{p}$.*

Proof. Take $S = A - \mathfrak{p}$ in (3.11) (iv). ∎

Remark. Thus the passage from A to $A_\mathfrak{p}$ cuts out all prime ideals except those contained in $\mathfrak{p}$. In the other direction, the passage from A to $A/\mathfrak{p}$ cuts out all prime ideals except those containing $\mathfrak{p}$. Hence if $\mathfrak{p}, \mathfrak{q}$ are prime ideals such that $\mathfrak{p} \supseteq \mathfrak{q}$, then by localizing with respect to $\mathfrak{p}$ and taking the quotient mod $\mathfrak{q}$ (in either order: these two operations commute, by (3.4)), we restrict our attention to those prime ideals which lie between $\mathfrak{p}$ and $\mathfrak{q}$. In particular, if $\mathfrak{p} = \mathfrak{q}$ we end up with a field, called the *residue field at* $\mathfrak{p}$, which can be obtained either as the field of fractions of the integral domain $A/\mathfrak{p}$ or as the residue field of the local ring $A_{\mathfrak{p}}$.

Proposition 3.14. *Let M be a finitely generated A-module, S a multiplicatively closed subset of A. Then $S^{-1}(\text{Ann}(M)) = \text{Ann}(S^{-1}M)$.*

Proof. If this is true for two A-modules, M, N, it is true for $M + N$:

$$\begin{aligned} S^{-1}(\text{Ann}(M+N)) &= S^{-1}(\text{Ann}(M) \cap \text{Ann}(N)) \text{ by (2.2)} \\ &= S^{-1}(\text{Ann}(M)) \cap S^{-1}(\text{Ann}(N)) \text{ by (3.4)} \\ &= \text{Ann}(S^{-1}M) \cap \text{Ann}(S^{-1}N) \text{ by hypothesis} \\ &= \text{Ann}(S^{-1}M + S^{-1}N) = \text{Ann}\big(S^{-1}(M+N)\big). \end{aligned}$$

Hence it is enough to prove (3.14) for M generated by a single element: then $M \cong A/\mathfrak{a}$ (as A-module), where $\mathfrak{a} = \text{Ann}(M)$; $S^{-1}M \cong (S^{-1}A)/(S^{-1}\mathfrak{a})$ by (3.4), so that $\text{Ann}(S^{-1}M) = S^{-1}\mathfrak{a} = S^{-1}(\text{Ann}(M))$. ∎

Corollary 3.15. *If N, P are submodules of an A-module M and if P is finitely generated, then $S^{-1}(N:P) = (S^{-1}N:S^{-1}P)$.*

Proof. $(N:P) = \text{Ann}\big((N+P)/N\big)$ by (2.2); now apply (3.14). ∎

Proposition 3.16. *Let $A \to B$ be a ring homomorphism and let $\mathfrak{p}$ be a prime ideal of A. Then $\mathfrak{p}$ is the contraction of a prime ideal of B if and only if $\mathfrak{p}^{ec} = \mathfrak{p}$.*

Proof. If $\mathfrak{p} = \mathfrak{q}^c$ then $\mathfrak{p}^{ec} = \mathfrak{p}$ by (1.17). Conversely, if $\mathfrak{p}^{ec} = \mathfrak{p}$, let S be the image of $A - \mathfrak{p}$ in B. Then $\mathfrak{p}^e$ does not meet S, therefore by (3.11) its extension in $S^{-1}B$ is a proper ideal and hence is contained in a maximal ideal $\mathfrak{m}$ of $S^{-1}B$. If $\mathfrak{q}$ is the contraction of $\mathfrak{m}$ in B, then $\mathfrak{q}$ is prime, $\mathfrak{q} \supseteq \mathfrak{p}^e$ and $\mathfrak{q} \cap S = \varnothing$. Hence $\mathfrak{q}^c = \mathfrak{p}$. ∎

EXERCISES

1. Let S be a multiplicatively closed subset of a ring A, and let M be a finitely generated A-module. Prove that $S^{-1}M = 0$ if and only if there exists $s \in S$ such that $sM = 0$.

2. Let $\mathfrak{a}$ be an ideal of a ring A, and let $S = 1 + \mathfrak{a}$. Show that $S^{-1}\mathfrak{a}$ is contained in the Jacobson radical of $S^{-1}A$.

   Use this result and Nakayama's lemma to give a proof of (2.5) which does not depend on determinants. [If $M = \mathfrak{a}M$, then $S^{-1}M = (S^{-1}\mathfrak{a})(S^{-1}M)$, hence by Nakayama we have $S^{-1}M = 0$. Now use Exercise 1.]

3. Let A be a ring, let S and T be two multiplicatively closed subsets of A, and let U be the image of T in $S^{-1}A$. Show that the rings $(ST)^{-1}A$ and $U^{-1}(S^{-1}A)$ are isomorphic.

4. Let $f: A \to B$ be a homomorphism of rings and let S be a multiplicatively closed subset of A. Let $T = f(S)$. Show that $S^{-1}B$ and $T^{-1}B$ are isomorphic as $S^{-1}A$-modules.

5. Let A be a ring. Suppose that, for each prime ideal $\mathfrak{p}$, the local ring $A_\mathfrak{p}$ has no nilpotent element $\neq 0$. Show that A has no nilpotent element $\neq 0$. If each $A_\mathfrak{p}$ is an integral domain, is A necessarily an integral domain?

6. Let A be a ring $\neq 0$ and let Σ be the set of all multiplicatively closed subsets S of A such that $0 \notin S$. Show that Σ has maximal elements, and that $S \in \Sigma$ is maximal if and only if $A - S$ is a minimal prime ideal of A.

7. A multiplicatively closed subset S of a ring A is said to be *saturated* if

$$xy \in S \Leftrightarrow x \in S \text{ and } y \in S.$$

Prove that

i) S is saturated $\Leftrightarrow A - S$ is a union of prime ideals.

ii) If S is any multiplicatively closed subset of A, there is a unique smallest saturated multiplicatively closed subset $\bar{S}$ containing S, and that $\bar{S}$ is the complement in A of the union of the prime ideals which do not meet S. ($\bar{S}$ is called the *saturation* of S.)

If $S = 1 + \mathfrak{a}$, where $\mathfrak{a}$ is an ideal of A, find $\bar{S}$.

8. Let S, T be multiplicatively closed subsets of A, such that $S \subseteq T$. Let $\phi: S^{-1}A \to T^{-1}A$ be the homomorphism which maps each $a/s \in S^{-1}A$ to a/s considered as an element of $T^{-1}A$. Show that the following statements are equivalent:

i) ϕ is bijective.

ii) For each $t \in T$, $t/1$ is a unit in $S^{-1}A$.

iii) For each $t \in T$ there exists $x \in A$ such that $xt \in S$.

iv) T is contained in the saturation of S (Exercise 7).

v) Every prime ideal which meets T also meets S.

9. The set S_0 of all non-zero-divisors in A is a saturated multiplicatively closed subset of A. Hence the set D of zero-divisors in A is a union of prime ideals (see Chapter 1, Exercise 14). Show that every minimal prime ideal of A is contained in D. [Use Exercise 6.]

The ring $S_0^{-1}A$ is called the *total ring of fractions* of A. Prove that

i) S_0 is the largest multiplicatively closed subset of A for which the homomorphism $A \to S_0^{-1}A$ is injective.

ii) Every element in $S_0^{-1}A$ is either a zero-divisor or a unit.

iii) Every ring in which every non-unit is a zero-divisor is equal to its total ring of fractions (i.e., $A \to S_0^{-1}A$ is bijective).

10. Let A be a ring.

i) If A is absolutely flat (Chapter 2, Exercise 27) and S is any multiplicatively closed subset of A, then $S^{-1}A$ is absolutely flat.

ii) A is absolutely flat $\Leftrightarrow A_\mathfrak{m}$ is a field for each maximal ideal $\mathfrak{m}$.

11. Let A be a ring. Prove that the following are equivalent:

i) $A/\mathfrak{N}$ is absolutely flat ($\mathfrak{N}$ being the nilradical of A).

ii) Every prime ideal of A is maximal.

iii) Spec (A) is a T_1-space (i.e., every subset consisting of a single point is closed).
iv) Spec (A) is Hausdorff.

If these conditions are satisfied, show that Spec (A) is compact and totally disconnected (i.e. the only connected subsets of Spec (A) are those consisting of a single point).

12. Let A be an integral domain and M an A-module. An element $x \in M$ is a *torsion element* of M if Ann $(x) \neq 0$, that is if x is killed by some non-zero element of A. Show that the torsion elements of M form a submodule of M. This submodule is called the *torsion submodule* of M and is denoted by $T(M)$. If $T(M) = 0$, the module M is said to be torsion-free. Show that
i) If M is any A-module, then $M/T(M)$ is torsion-free.
ii) If $f: M \to N$ is a module homomorphism, then $f(T(M)) \subseteq T(N)$.
iii) If $0 \to M' \to M \to M''$ is an exact sequence, then the sequence $0 \to T(M') \to T(M) \to T(M'')$ is exact.
iv) If M is any A-module, then $T(M)$ is the kernel of the mapping $x \mapsto 1 \otimes x$ of M into $K \otimes_A M$, where K is the field of fractions of A.
[For iv), show that K may be regarded as the direct limit of its submodules $A\xi$ ($\xi \in K$); using Chapter 1, Exercise 15 and Exercise 20, show that if $1 \otimes x = 0$ in $K \otimes M$ then $1 \otimes x = 0$ in $A\xi \otimes M$ for some $\xi \neq 0$. Deduce that $\xi^{-1}x = 0$.]

13. Let S be a multiplicatively closed subset of an integral domain A. In the notation of Exercise 12, show that $T(S^{-1}M) = S^{-1}(TM)$. Deduce that the following are equivalent:
i) M is torsion-free.
ii) $M_{\mathfrak{p}}$ is torsion-free for all prime ideals $\mathfrak{p}$.
iii) $M_{\mathfrak{m}}$ is torsion-free for all maximal ideals $\mathfrak{m}$.

14. Let M be an A-module and $\mathfrak{a}$ an ideal of A. Suppose that $M_{\mathfrak{m}} = 0$ for all maximal ideals $\mathfrak{m} \supseteq \mathfrak{a}$. Prove that $M = \mathfrak{a}M$. [Pass to the $A/\mathfrak{a}$-module $M/\mathfrak{a}M$ and use (3.8).]

15. Let A be a ring, and let F be the A-module A^n. Show that every set of n generators of F is a basis of F. [Let $x_1, \ldots, x_n$ be a set of generators and $e_1, \ldots, e_n$ the canonical basis of F. Define $\phi: F \to F$ by $\phi(e_i) = x_i$. Then ϕ is surjective and we have to prove that it is an isomorphism. By (3.9) we may assume that A is a local ring. Let N be the kernel of ϕ and let $k = A/\mathfrak{m}$ be the residue field of A. Since F is a flat A-module, the exact sequence $0 \to N \to F \to F \to 0$ gives an exact sequence $0 \longrightarrow k \otimes N \longrightarrow k \otimes F \xrightarrow{1 \otimes \phi} k \otimes F \longrightarrow 0$. Now $k \otimes F = k^n$ is an n-dimensional vector space over k; $1 \otimes \phi$ is surjective, hence bijective, hence $k \otimes N = 0$.

Also N is finitely generated, by Chapter 2, Exercise 12, hence $N = 0$ by Nakayama's lemma. Hence ϕ is an isomorphism.]

Deduce that every set of generators of F has at least n elements.

16. Let B be a flat A-algebra. Then the following conditions are equivalent:
i) $\mathfrak{a}^{ec} = \mathfrak{a}$ for all ideals $\mathfrak{a}$ of A.
ii) Spec $(B) \to$ Spec (A) is surjective.
iii) For every maximal ideal $\mathfrak{m}$ of A we have $\mathfrak{m}^e \neq (1)$.

iv) If M is any non-zero A-module, then $M_B \neq 0$.
v) For every A-module M, the mapping $x \mapsto 1 \otimes x$ of M into M_B is injective.
[For i) $\Rightarrow$ ii), use (3.16). ii) $\Rightarrow$ iii) is clear.
iii) $\Rightarrow$ iv): Let x be a non-zero element of M and let $M' = Ax$. Since B is flat over A it is enough to show that $M'_B \neq 0$. We have $M' \cong A/\mathfrak{a}$ for some ideal $\mathfrak{a} \neq (1)$, hence $M'_B \cong B/\mathfrak{a}^e$. Now $\mathfrak{a} \subseteq \mathfrak{m}$ for some maximal ideal $\mathfrak{m}$, hence $\mathfrak{a}^e \subseteq \mathfrak{m}^e \neq (1)$. Hence $M'_B \neq 0$.
iv) $\Rightarrow$ v): Let M' be the kernel of $M \to M_B$. Since B is flat over A, the sequence $0 \to M'_B \to M_B \to (M_B)_B$ is exact. But (Chapter 2, Exercise 13, with $N = M_B$) the mapping $M_B \to (M_B)_B$ is injective, hence $M'_B = 0$ and therefore $M' = 0$.
v) $\Rightarrow$ i): Take $M = A/\mathfrak{a}$.]
B is said to be *faithfully flat* over A.

17. Let $A \xrightarrow{f} B \xrightarrow{g} C$ be ring homomorphisms. If $g \circ f$ is flat and g is faithfully flat, then f is flat.

18. Let $f: A \to B$ be a flat homomorphism of rings, let $\mathfrak{q}$ be a prime ideal of B and let $\mathfrak{p} = \mathfrak{q}^c$. Then $f^*: \text{Spec}(B_\mathfrak{q}) \to \text{Spec}(A_\mathfrak{p})$ is surjective. [For $B_\mathfrak{p}$ is flat over $A_\mathfrak{p}$ by (3.10), and $B_\mathfrak{q}$ is a local ring of $B_\mathfrak{p}$, hence is flat over $B_\mathfrak{p}$. Hence $B_\mathfrak{q}$ is flat over $A_\mathfrak{p}$ and satisfies condition (3) of Exercise 16.]

19. Let A be a ring, M an A-module. The *support* of M is defined to be the set $\text{Supp}(M)$ of prime ideals $\mathfrak{p}$ of A such that $M_\mathfrak{p} \neq 0$. Prove the following results:
i) $M \neq 0 \Leftrightarrow \text{Supp}(M) \neq \varnothing$.
ii) $V(\mathfrak{a}) = \text{Supp}(A/\mathfrak{a})$.
iii) If $0 \to M' \to M \to M'' \to 0$ is an exact sequence, then $\text{Supp}(M) = \text{Supp}(M') \cup \text{Supp}(M'')$.
iv) If $M = \sum M_i$, then $\text{Supp}(M) = \bigcup \text{Supp}(M_i)$.
v) If M is finitely generated, then $\text{Supp}(M) = V(\text{Ann}(M))$ (and is therefore a closed subset of $\text{Spec}(A)$).
vi) If M, N are finitely generated, then $\text{Supp}(M \otimes_A N) = \text{Supp}(M) \cap \text{Supp}(N)$. [Use Chapter 2, Exercise 3.]
vii) If M is finitely generated and $\mathfrak{a}$ is an ideal of A, then $\text{Supp}(M/\mathfrak{a}M) = V(\mathfrak{a} + \text{Ann}(M))$.
viii) If $f: A \to B$ is a ring homomorphism and M is a finitely generated A-module, then $\text{Supp}(B \otimes_A M) = f^{*-1}(\text{Supp}(M))$.

20. Let $f: A \to B$ be a ring homomorphism, $f^*: \text{Spec}(B) \to \text{Spec}(A)$ the associated mapping. Show that
i) Every prime ideal of A is a contracted ideal $\Leftrightarrow f^*$ is surjective.
ii) Every prime ideal of B is an extended ideal $\Rightarrow f^*$ is injective.
Is the converse of ii) true?

21. i) Let A be a ring, S a multiplicatively closed subset of A, and $\phi: A \to S^{-1}A$ the canonical homomorphism. Show that $\phi^*: \text{Spec}(S^{-1}A) \to \text{Spec}(A)$ is a homeomorphism of $\text{Spec}(S^{-1}A)$ onto its image in $X = \text{Spec}(A)$. Let this image be denoted by $S^{-1}X$.
In particular, if $f \in A$, the image of $\text{Spec}(A_f)$ in X is the basic open set X_f (Chapter 1, Exercise 17).

ii) Let $f\colon A \to B$ be a ring homomorphism. Let $X = \operatorname{Spec}(A)$ and $Y = \operatorname{Spec}(B)$, and let $f^*\colon Y \to X$ be the mapping associated with f. Identifying $\operatorname{Spec}(S^{-1}A)$ with its canonical image $S^{-1}X$ in X, and $\operatorname{Spec}(S^{-1}B)$ $(= \operatorname{Spec}(f(S)^{-1}B))$ with its canonical image $S^{-1}Y$ in Y, show that $S^{-1}f^*\colon \operatorname{Spec}(S^{-1}B) \to \operatorname{Spec}(S^{-1}A)$ is the restriction of f^* to $S^{-1}Y$, and that $S^{-1}Y = f^{*-1}(S^{-1}X)$.

iii) Let $\mathfrak{a}$ be an ideal of A and let $\mathfrak{b} = \mathfrak{a}^e$ be its extension in B. Let $\bar{f}\colon A/\mathfrak{a} \to B/\mathfrak{b}$ be the homomorphism induced by f. If $\operatorname{Spec}(A/\mathfrak{a})$ is identified with its canonical image $V(\mathfrak{a})$ in X, and $\operatorname{Spec}(B/\mathfrak{b})$ with its image $V(\mathfrak{b})$ in Y, show that $\bar{f}^*$ is the restriction of f^* to $V(\mathfrak{b})$.

iv) Let $\mathfrak{p}$ be a prime ideal of A. Take $S = A - \mathfrak{p}$ in ii) and then reduce mod $S^{-1}\mathfrak{p}$ as in iii). Deduce that the subspace $f^{*-1}(\mathfrak{p})$ of Y is naturally homeomorphic to $\operatorname{Spec}(B_\mathfrak{p}/\mathfrak{p}B_\mathfrak{p}) = \operatorname{Spec}(k(\mathfrak{p}) \otimes_A B)$, where $k(\mathfrak{p})$ is the residue field of the local ring $A_\mathfrak{p}$.

$\operatorname{Spec}(k(\mathfrak{p}) \otimes_A B)$ is called the *fiber* of f^* over $\mathfrak{p}$.

22. Let A be a ring and $\mathfrak{p}$ a prime ideal of A. Then the canonical image of $\operatorname{Spec}(A_\mathfrak{p})$ in $\operatorname{Spec}(A)$ is equal to the intersection of all the open neighborhoods of $\mathfrak{p}$ in $\operatorname{Spec}(A)$.

23. Let A be a ring, let $X = \operatorname{Spec}(A)$ and let U be a basic open set in X (i.e., $U = X_f$ for some $f \in A$: Chapter 1, Exercise 17).

i) If $U = X_f$, show that the ring $A(U) = A_f$ depends only on U and not on f.

ii) Let $U' = X_g$ be another basic open set such that $U' \subseteq U$. Show that there is an equation of the form $g^n = uf$ for some integer $n > 0$ and some $u \in A$, and use this to define a homomorphism $\rho\colon A(U) \to A(U')$ (i.e., $A_f \to A_g$) by mapping a/f^m to au^m/g^{mn}. Show that ρ depends only on U and U'. This homomorphism is called the *restriction* homomorphism.

iii) If $U = U'$, then ρ is the identity map.

iv) If $U \supseteq U' \supseteq U''$ are basic open sets in X, show that the diagram

$$\begin{array}{ccc} A(U) & \longrightarrow & A(U'') \\ & \searrow \quad \nearrow & \\ & A(U') & \end{array}$$

(in which the arrows are restriction homomorphisms) is commutative.

v) Let $x\ (= \mathfrak{p})$ be a point of X. Show that

$$\varinjlim_{U \ni x} A(U) \cong A_\mathfrak{p}.$$

The assignment of the ring $A(U)$ to each basic open set U of X, and the restriction homomorphisms ρ, satisfying the conditions iii) and iv) above, constitutes a *presheaf of rings* on the basis of open sets $(X_f)_{f \in A}$. v) says that the stalk of this presheaf at $x \in X$ is the corresponding local ring $A_\mathfrak{p}$.

24. Show that the presheaf of Exercise 23 has the following property. Let $(U_i)_{i \in I}$ be a covering of X by basic open sets. For each $i \in I$ let $s_i \in A(U_i)$ be such that, for each pair of indices i, j, the images of s_i and s_j in $A(U_i \cap U_j)$ are equal. Then there exists a unique $s \in A\ (= A(X))$ whose image in $A(U_i)$ is s_i, for all $i \in I$. (This essentially implies that the presheaf is a *sheaf*.)

25. Let $f: A \to B$, $g: A \to C$ be ring homomorphisms and let $h: A \to B \otimes_A C$ be defined by $h(x) = f(x) \otimes g(x)$. Let X, Y, Z, T be the prime spectra of A, B, C, $B \otimes_A C$ respectively. Then $h^*(T) = f^*Y \cap g^*(Z)$.
[Let $\mathfrak{p} \in X$, and let $k = k(\mathfrak{p})$ be the residue field at $\mathfrak{p}$. By Exercise 21, the fiber $h^{*-1}(\mathfrak{p})$ is the spectrum of $(B \otimes_A C) \otimes_A k \cong (B \otimes_A k) \otimes_k (C \otimes_A k)$. Hence $\mathfrak{p} \in h^*(T) \Leftrightarrow (B \otimes_A k) \otimes_k (C \otimes_A k) \neq 0 \Leftrightarrow B \otimes_A k \neq 0$ and $C \otimes_A k \neq 0 \Leftrightarrow \mathfrak{p} \in f^*(Y) \cap g^*(Z)$.]

26. Let $(B_\alpha, g_{\alpha\beta})$ be a direct system of rings and B the direct limit. For each α, let $f_\alpha: A \to B_\alpha$ be a ring homomorphism such that $g_{\alpha\beta} \circ f_\alpha = f_\beta$ whenever $\alpha \leqslant \beta$ (i.e. the B_α form a direct system of A-algebras). The f_α induce $f: A \to B$. Show that

$$f^*(\operatorname{Spec}(B)) = \bigcap_\alpha f_\alpha^*(\operatorname{Spec}(B_\alpha)).$$

[Let $\mathfrak{p} \in \operatorname{Spec}(A)$. Then $f^{*-1}(\mathfrak{p})$ is the spectrum of

$$B \otimes_A k(\mathfrak{p}) \cong \varinjlim (B_\alpha \otimes_A k(\mathfrak{p}))$$

(since tensor products commute with direct limits: Chapter 2, Exercise 20). By Exercise 21 of Chapter 2 it follows that $f^{*-1}(\mathfrak{p}) = \varnothing$ if and only if $B_\alpha \otimes_A k(\mathfrak{p}) = 0$ for some α, i.e., if and only if $f_\alpha^{*-1}(\mathfrak{p}) = \varnothing$.]

27. i) Let $f_\alpha: A \to B_\alpha$ be any family of A-algebras and let $f: A \to B$ be their tensor product over A (Chapter 2, Exercise 23). Then

$$f^*(\operatorname{Spec}(B)) = \bigcap_\alpha f_\alpha^*(\operatorname{Spec}(B_\alpha)).$$

[Use Examples 25 and 26.]

ii) Let $f_\alpha: A \to B_\alpha$ be any finite family of A-algebras and let $B = \prod_\alpha B_\alpha$. Define $f: A \to B$ by $f(x) = (f_\alpha(x))$. Then $f^*(\operatorname{Spec}(B)) = \bigcup_\alpha f_\alpha^*(\operatorname{Spec}(B_\alpha))$.

iii) Hence the subsets of $X = \operatorname{Spec}(A)$ of the form $f^*(\operatorname{Spec}(B))$, where $f: A \to B$ is a ring homomorphism, satisfy the axioms for closed sets in a topological space. The associated topology is the *constructible* topology on X. It is finer than the Zariski topology (i.e., there are more open sets, or equivalently more closed sets).

iv) Let X_C denote the set X endowed with the constructible topology. Show that X_C is quasi-compact.

28. (Continuation of Exercise 27.)

i) For each $g \in A$, the set X_g (Chapter 1, Exercise 17) is both open and closed in the constructible topology.

ii) Let C' denote the smallest topology on X for which the sets X_g are both open and closed, and let $X_{C'}$ denote the set X endowed with this topology. Show that $X_{C'}$ is Hausdorff.

iii) Deduce that the identity mapping $X_C \to X_{C'}$ is a homeomorphism. Hence a subset E of X is of the form $f^*(\operatorname{Spec}(B))$ for some $f: A \to B$ if and only if it is closed in the topology C'.

iv) The topological space X_C is compact, Hausdorff and totally disconnected.

29. Let $f: A \to B$ be a ring homomorphism. Show that $f^*: \text{Spec}(B) \to \text{Spec}(A)$ is a continuous *closed* mapping (i.e., maps closed sets to closed sets) for the constructible topology.

30. Show that the Zariski topology and the constructible topology on Spec (A) are the same if and only if $A/\mathfrak{N}$ is absolutely flat (where $\mathfrak{N}$ is the nilradical of A). [Use Exercise 11.]

4

Primary Decomposition

The decomposition of an ideal into primary ideals is a traditional pillar of ideal theory. It provides the algebraic foundation for decomposing an algebraic variety into its irreducible components—although it is only fair to point out that the algebraic picture is more complicated than naïve geometry would suggest. From another point of view primary decomposition provides a generalization of the factorization of an integer as a product of prime-powers. In the modern treatment, with its emphasis on localization, primary decomposition is no longer such a central tool in the theory. It is still, however, of interest in itself and in this chapter we establish the classical uniqueness theorems.

The prototypes of commutative rings are **Z** and the ring of polynomials $k[x_1, \ldots, x_n]$ where k is a field; both these are unique factorization domains. This is not true of arbitrary commutative rings, even if they are integral domains (the classical example is the ring $\mathbf{Z}[\sqrt{-5}]$, in which the element 6 has two essentially distinct factorizations, $2 \cdot 3$ and $(1 + \sqrt{-5})(1 - \sqrt{-5})$). However, there is a generalized form of "unique factorization" of *ideals* (not of elements) in a wide class of rings (the Noetherian rings).

A prime ideal in a ring A is in some sense a generalization of a prime number. The corresponding generalization of a power of a prime number is a primary ideal. An ideal $\mathfrak{q}$ in a ring A is *primary* if $\mathfrak{q} \neq A$ and if

$$xy \in \mathfrak{q} \Rightarrow \text{either } x \in \mathfrak{q} \text{ or } y^n \in \mathfrak{q} \text{ for some } n > 0.$$

In other words,

$$\mathfrak{q} \text{ is primary} \Leftrightarrow A/\mathfrak{q} \neq 0 \text{ and every zero-divisor in } A/\mathfrak{q} \text{ is nilpotent.}$$

Clearly every prime ideal is primary. Also the contraction of a primary ideal is primary, for if $f: A \to B$ and if $\mathfrak{q}$ is a primary ideal in B, then $A/\mathfrak{q}^c$ is isomorphic to a subring of $B/\mathfrak{q}$.

Proposition 4.1. *Let $\mathfrak{q}$ be a primary ideal in a ring A. Then $r(\mathfrak{q})$ is the smallest prime ideal containing $\mathfrak{q}$.*

Proof. By (1.8) it is enough to show that $\mathfrak{p} = r(\mathfrak{q})$ is prime. Let $xy \in r(\mathfrak{q})$, then $(xy)^m \in \mathfrak{q}$ for some $m > 0$, and therefore either $x^m \in \mathfrak{q}$ or $y^{mn} \in \mathfrak{q}$ for some $n > 0$; i.e., either $x \in r(\mathfrak{q})$ or $y \in r(\mathfrak{q})$. ∎

If $\mathfrak{p} = r(\mathfrak{q})$, then $\mathfrak{q}$ is said to be $\mathfrak{p}$-*primary*.

Examples. 1) The primary ideals in $\mathbf{Z}$ are (0) and (p^n), where p is prime. For these are the only ideals in $\mathbf{Z}$ with prime radical, and it is immediately checked that they are primary.

2) Let $A = k[x, y]$, $\mathfrak{q} = (x, y^2)$. Then $A/\mathfrak{q} \cong k[y]/(y^2)$, in which the zero-divisors are all the multiples of y, hence are nilpotent. Hence $\mathfrak{q}$ is primary, and its radical $\mathfrak{p}$ is (x, y). We have $\mathfrak{p}^2 \subset \mathfrak{q} \subset \mathfrak{p}$ (strict inclusions), so that a primary ideal is not necessarily a prime-power.

3) Conversely, a prime power $\mathfrak{p}^n$ is not necessarily primary, although its radical is the prime ideal $\mathfrak{p}$. For example, let $A = k[x, y, z]/(xy - z^2)$ and let $\bar{x}, \bar{y}, \bar{z}$ denote the images of x, y, z respectively in A. Then $\mathfrak{p} = (\bar{x}, \bar{z})$ is prime (since $A/\mathfrak{p} \cong k[y]$, an integral domain); we have $\bar{x}\bar{y} = \bar{z}^2 \in \mathfrak{p}^2$ but $\bar{x} \notin \mathfrak{p}^2$ and $\bar{y} \notin r(\mathfrak{p}^2) = \mathfrak{p}$; hence $\mathfrak{p}^2$ is not primary. However, there is the following result:

Proposition 4.2. *If $r(\mathfrak{a})$ is maximal, then $\mathfrak{a}$ is primary. In particular, the powers of a maximal ideal $\mathfrak{m}$ are $\mathfrak{m}$-primary.*

Proof. Let $r(\mathfrak{a}) = \mathfrak{m}$. The image of $\mathfrak{m}$ in $A/\mathfrak{a}$ is the nilradical of $A/\mathfrak{a}$, hence $A/\mathfrak{a}$ has only one prime ideal, by (1.8). Hence every element of $A/\mathfrak{a}$ is either a unit or nilpotent, and so every zero-divisor in $A/\mathfrak{a}$ is nilpotent. ■

We are going to study presentations of an ideal as an *intersection of primary ideals*. First, a couple of lemmas:

Lemma 4.3. *If $\mathfrak{q}_i$ $(1 \leqslant i \leqslant n)$ are $\mathfrak{p}$-primary, then $\mathfrak{q} = \bigcap_{i=1}^n \mathfrak{q}_i$ is $\mathfrak{p}$-primary.*

Proof. $r(\mathfrak{q}) = r(\bigcap_{i=1}^n \mathfrak{q}_i) = \bigcap r(\mathfrak{q}_i) = \mathfrak{p}$. Let $xy \in \mathfrak{q}$, $y \notin \mathfrak{q}$. Then for some i we have $xy \in \mathfrak{q}_i$ and $y \notin \mathfrak{q}_i$, hence $x \in \mathfrak{p}$, since $\mathfrak{q}_i$ is primary. ■

Lemma 4.4. *Let $\mathfrak{q}$ be a $\mathfrak{p}$-primary ideal, x an element of A. Then*

i) *if $x \in \mathfrak{q}$ then $(\mathfrak{q}:x) = (1)$;*

ii) *if $x \notin \mathfrak{q}$ then $(\mathfrak{q}:x)$ is $\mathfrak{p}$-primary, and therefore $r(\mathfrak{q}:x) = \mathfrak{p}$;*

iii) *if $x \notin \mathfrak{p}$ then $(\mathfrak{q}:x) = \mathfrak{q}$.*

Proof. i) and iii) follow immediately from the definitions.

ii): if $y \in (\mathfrak{q}:x)$ then $xy \in \mathfrak{q}$, hence (as $x \notin \mathfrak{q}$) we have $y \in \mathfrak{p}$. Hence $\mathfrak{q} \subseteq (\mathfrak{q}:x) \subseteq \mathfrak{p}$; taking radicals, we get $r(\mathfrak{q}:x) = \mathfrak{p}$. Let $yz \in (\mathfrak{q}:x)$ with $y \notin \mathfrak{p}$; then $xyz \in \mathfrak{q}$, hence $xz \in \mathfrak{q}$, hence $z \in (\mathfrak{q}:x)$. ■

A *primary decomposition* of an ideal $\mathfrak{a}$ in A is an expression of $\mathfrak{a}$ as a finite intersection of primary ideals, say

$$\mathfrak{a} = \bigcap_{i=1}^{n} \mathfrak{q}_i. \tag{1}$$

(In general such a primary decomposition need not exist; in this chapter we shall restrict our attention to ideals which have a primary decomposition.) If moreover (i) the $r(\mathfrak{q}_i)$ are all distinct, and (ii) we have $\mathfrak{q}_i \not\supseteq \bigcap_{j\neq i} \mathfrak{q}_j$ $(1 \leqslant i \leqslant n)$ the primary decomposition (1) is said to be *minimal* (or irredundant, or reduced, or normal, ...). By (4.3) we can achieve (i) and then we can omit any superfluous terms to achieve (ii); thus any primary decomposition can be reduced to a minimal one. We shall say that $\mathfrak{a}$ is *decomposable* if it has a primary decomposition.

Theorem 4.5. (1st uniqueness theorem). *Let $\mathfrak{a}$ be a decomposable ideal and let $\mathfrak{a} = \bigcap_{i=1}^n \mathfrak{q}_i$ be a minimal primary decomposition of $\mathfrak{a}$. Let $\mathfrak{p}_i = r(\mathfrak{q}_i)$ $(1 \leqslant i \leqslant n)$. Then the $\mathfrak{p}_i$ are precisely the prime ideals which occur in the set of ideals $r(\mathfrak{a}:x)$ $(x \in A)$, and hence are independent of the particular decomposition of $\mathfrak{a}$.*

Proof. For any $x \in A$ we have $(\mathfrak{a}:x) = (\bigcap \mathfrak{q}_i : x) = \bigcap (\mathfrak{q}_i : x)$, hence $r(\mathfrak{a}:x) = \bigcap_{i=1}^n r(\mathfrak{q}_i:x) = \bigcap_{x \notin \mathfrak{q}_j} \mathfrak{p}_j$ by (4.4). Suppose $r(\mathfrak{a}:x)$ is prime; then by (1.11) we have $r(\mathfrak{a}:x) = \mathfrak{p}_j$ for some j. Hence every prime ideal of the form $r(\mathfrak{a}:x)$ is one of the $\mathfrak{p}_j$. Conversely, for each i there exists $x_i \notin \mathfrak{q}_i$, $x_i \in \bigcap_{j\neq i} \mathfrak{q}_j$, since the decomposition is minimal; and we have $r(\mathfrak{a}:x_i) = \mathfrak{p}_i$. ∎

Remarks. 1) The above proof, coupled with the last part of (4.4), shows that for each i there exists x_i in A such that $(\mathfrak{a}:x_i)$ is $\mathfrak{p}_i$-primary.

2) Considering $A/\mathfrak{a}$ as an A-module, (4.5) is equivalent to saying that the $\mathfrak{p}_i$ are precisely the prime ideals which occur as radicals of annihilators of elements of $A/\mathfrak{a}$.

Example. Let $\mathfrak{a} = (x^2, xy)$ in $A = k[x, y]$. Then $\mathfrak{a} = \mathfrak{p}_1 \cap \mathfrak{p}_2^2$ where $\mathfrak{p}_1 = (x)$, $\mathfrak{p}_2 = (x, y)$. The ideal $\mathfrak{p}_2^2$ is primary by (4.2). So the prime ideals are $\mathfrak{p}_1, \mathfrak{p}_2$. In this example $\mathfrak{p}_1 \subset \mathfrak{p}_2$; we have $r(\mathfrak{a}) = \mathfrak{p}_1 \cap \mathfrak{p}_2 = \mathfrak{p}_1$, but $\mathfrak{a}$ is not a primary ideal.

The prime ideals $\mathfrak{p}_i$ in (4.5) are said to *belong* to $\mathfrak{a}$, or to be *associated* with $\mathfrak{a}$. The ideal $\mathfrak{a}$ is primary if and only if it has only one associated prime ideal. The minimal elements of the set $\{\mathfrak{p}_1, \ldots, \mathfrak{p}_n\}$ are called the *minimal* or *isolated* prime ideals belonging to $\mathfrak{a}$. The others are called *embedded* prime ideals. In the example above, $\mathfrak{p}_2 = (x, y)$ is embedded.

Proposition 4.6. *Let $\mathfrak{a}$ be a decomposable ideal. Then any prime ideal $\mathfrak{p} \supseteq \mathfrak{a}$ contains a minimal prime ideal belonging to $\mathfrak{a}$, and thus the minimal prime ideals of $\mathfrak{a}$ are precisely the minimal elements in the set of all prime ideals containing $\mathfrak{a}$.*

Proof. If $\mathfrak{p} \supseteq \mathfrak{a} = \bigcap_{i=1}^n \mathfrak{q}_i$, then $\mathfrak{p} = r(\mathfrak{p}) \supseteq \bigcap r(\mathfrak{q}_i) = \bigcap \mathfrak{p}_i$. Hence by (1.11) we have $\mathfrak{p} \supseteq \mathfrak{p}_i$ for some i; hence $\mathfrak{p}$ contains a minimal prime ideal of $\mathfrak{a}$. ∎

Remarks. 1) The names *isolated* and *embedded* come from geometry. Thus if $A = k[x_1, \ldots, x_n]$ where k is a field, the ideal $\mathfrak{a}$ gives rise to a variety $X \subseteq k^n$ (see Chapter 1, Exercise 25). The minimal primes $\mathfrak{p}_i$ correspond to the irreducible components of X, and the embedded primes correspond to subvarieties of these, i.e., varieties *embedded* in the irreducible components. Thus in the example before (4.6) the variety defined by $\mathfrak{a}$ is the line $x = 0$, and the embedded ideal $\mathfrak{p}_2 = (x, y)$ corresponds to the origin $(0, 0)$.

2) It is *not* true that all the primary components are independent of the decomposition. For example $(x^2, xy) = (x) \cap (x, y)^2 = (x) \cap (x^2, y)$ are two distinct minimal primary decompositions. However, there are some uniqueness properties: see (4.10).

***Proposition* 4.7.** *Let $\mathfrak{a}$ be a decomposable ideal, let $\mathfrak{a} = \bigcap_{i=1}^n \mathfrak{q}_i$ be a minimal primary decomposition, and let $r(\mathfrak{q}_i) = \mathfrak{p}_i$. Then*

$$\bigcup_{i=1}^{n} \mathfrak{p}_i = \{x \in A : (\mathfrak{a}:x) \neq \mathfrak{a}\}.$$

In particular, if the zero ideal is decomposable, the set D of zero-divisors of A is the union of the prime ideals belonging to 0.

Proof. If $\mathfrak{a}$ is decomposable, then 0 is decomposable in $A/\mathfrak{a}$: namely $0 = \bigcap \bar{\mathfrak{q}}_i$ where $\bar{\mathfrak{q}}_i$ is the image of $\mathfrak{q}_i$ in $A/\mathfrak{a}$, and is primary. Hence it is enough to prove the last statement of (4.7). By (1.15) we have $D = \bigcup_{x \neq 0} r(0:x)$; from the proof of (4.5), we have $r(0:x) = \bigcap_{x \notin \mathfrak{q}_j} \mathfrak{p}_j \subseteq \mathfrak{p}_j$ for some j, hence $D \subseteq \bigcup_{i=1}^n \mathfrak{p}_i$. But also from (4.5) each $\mathfrak{p}_i$ is of the form $r(0:x)$ for some $x \in A$, hence $\bigcup \mathfrak{p}_i \subseteq D$. ∎

Thus (the zero ideal being decomposable)

$$\begin{aligned} D &= \text{set of zero-divisors} \\ &= \textstyle\bigcup \text{ of all prime ideals belonging to } 0; \\ \mathfrak{N} &= \text{set of nilpotent elements} \\ &= \textstyle\bigcap \text{ of all minimal primes belonging to } 0. \end{aligned}$$

Next we investigate the behavior of primary ideals under localization.

***Proposition* 4.8.** *Let S be a multiplicatively closed subset of A, and let $\mathfrak{q}$ be a $\mathfrak{p}$-primary ideal.*

i) *If $S \cap \mathfrak{p} \neq \varnothing$, then $S^{-1}\mathfrak{q} = S^{-1}A$.*

ii) *If $S \cap \mathfrak{p} = \varnothing$, then $S^{-1}\mathfrak{q}$ is $S^{-1}\mathfrak{p}$-primary and its contraction in A is $\mathfrak{q}$.*

Hence primary ideals correspond to primary ideals in the correspondence (3.11) *between ideals in $S^{-1}A$ and contracted ideals in A.*

Proof. i) If $s \in S \cap \mathfrak{p}$, then $s^n \in S \cap \mathfrak{q}$ for some $n > 0$; hence $S^{-1}\mathfrak{q}$ contains $s^n/1$, which is a unit in $S^{-1}A$.

ii) If $S \cap \mathfrak{p} = \varnothing$, then $s \in S$ and $as \in \mathfrak{q}$ imply $a \in \mathfrak{q}$, hence $\mathfrak{q}^{ec} = \mathfrak{q}$ by (3.11). Also from (3.11) we have $r(\mathfrak{q}^e) = r(S^{-1}\mathfrak{q}) = S^{-1}r(\mathfrak{q}) = S^{-1}\mathfrak{p}$. The verification that $S^{-1}\mathfrak{q}$ is primary is straightforward. Finally, the contraction of a primary ideal is primary. ∎

For any ideal $\mathfrak{a}$ and any multiplicatively closed subset S in A, the contraction in A of the ideal $S^{-1}\mathfrak{a}$ is denoted by $S(\mathfrak{a})$.

Proposition 4.9. *Let S be a multiplicatively closed subset of A and let $\mathfrak{a}$ be a decomposable ideal. Let $\mathfrak{a} = \bigcap_{i=1}^{n} \mathfrak{q}_i$ be a minimal primary decomposition of $\mathfrak{a}$. Let $\mathfrak{p}_i = r(\mathfrak{q}_i)$ and suppose the $\mathfrak{q}_i$ numbered so that S meets $\mathfrak{p}_{m+1}, \ldots, \mathfrak{p}_n$ but not $\mathfrak{p}_1, \ldots, \mathfrak{p}_m$. Then*

$$S^{-1}\mathfrak{a} = \bigcap_{i=1}^{m} S^{-1}\mathfrak{q}_i, \qquad S(\mathfrak{a}) = \bigcap_{i=1}^{m} \mathfrak{q}_i,$$

and these are minimal primary decompositions.

Proof. $S^{-1}\mathfrak{a} = \bigcap_{i=1}^{n} S^{-1}\mathfrak{q}_i$ by (3.11) $= \bigcap_{i=1}^{m} S^{-1}\mathfrak{q}_i$ by (4.8), and $S^{-1}\mathfrak{q}_i$ is $S^{-1}\mathfrak{p}_i$-primary for $i = 1, \ldots, m$. Since the $\mathfrak{p}_i$ are distinct, so are the $S^{-1}\mathfrak{p}_i$ $(1 \leqslant i \leqslant m)$, hence we have a minimal primary decomposition. Contracting both sides, we get

$$S(\mathfrak{a}) = (S^{-1}\mathfrak{a})^c = \bigcap_{i=1}^{m} (S^{-1}\mathfrak{q}_i)^c = \bigcap_{i=1}^{m} \mathfrak{q}_i$$

by (4.8) again. ■

A set Σ of prime ideals belonging to $\mathfrak{a}$ is said to be *isolated* if it satisfies the following condition: if $\mathfrak{p}'$ is a prime ideal belonging to $\mathfrak{a}$ and $\mathfrak{p}' \subseteq \mathfrak{p}$ for some $\mathfrak{p} \in \Sigma$, then $\mathfrak{p}' \in \Sigma$.

Let Σ be an isolated set of prime ideals belonging to $\mathfrak{a}$, and let $S = A - \bigcup_{\mathfrak{p} \in \Sigma} \mathfrak{p}$. Then S is multiplicatively closed and, for any prime ideal $\mathfrak{p}'$ belonging to $\mathfrak{a}$, we have

$$\mathfrak{p}' \in \Sigma \Rightarrow \mathfrak{p}' \cap S = \varnothing;$$

$$\mathfrak{p}' \notin \Sigma \Rightarrow \mathfrak{p}' \nsubseteq \bigcup_{\mathfrak{p} \in \Sigma} \mathfrak{p} \text{ (by (1.11))} \Rightarrow \mathfrak{p}' \cap S \neq \varnothing.$$

Hence, from (4.9), we deduce

Theorem 4.10. (2nd uniqueness theorem). *Let $\mathfrak{a}$ be a decomposable ideal, let $\mathfrak{a} = \bigcap_{i=1}^{n} \mathfrak{q}_i$ be a minimal primary decomposition of $\mathfrak{a}$, and let $\{\mathfrak{p}_{i_1}, \ldots, \mathfrak{p}_{i_m}\}$ be an isolated set of prime ideals of $\mathfrak{a}$. Then $\mathfrak{q}_{i_1} \cap \cdots \cap \mathfrak{q}_{i_m}$ is independent of the decomposition.*

In particular:

Corollary 4.11. *The isolated primary components (i.e., the primary components $\mathfrak{q}_i$ corresponding to minimal prime ideals $\mathfrak{p}_i$) are uniquely determined by $\mathfrak{a}$.*

Proof of (4.10). We have $\mathfrak{q}_{i_1} \cap \cdots \cap \mathfrak{q}_{i_m} = S(\mathfrak{a})$ where $S = A - \mathfrak{p}_{i_1} \cup \cdots \cup \mathfrak{p}_{i_m}$, hence depends only on $\mathfrak{a}$ (since the $\mathfrak{p}_i$ depend only on $\mathfrak{a}$). ■

Remark. On the other hand, the embedded primary components are in general not uniquely determined by $\mathfrak{a}$. If A is a Noetherian ring, there are in fact infinitely many choices for each embedded component (see Chapter 8, Exercise 1).

EXERCISES

1. If an ideal $\mathfrak{a}$ has a primary decomposition, then Spec $(A/\mathfrak{a})$ has only finitely many irreducible components.
2. If $\mathfrak{a} = r(\mathfrak{a})$, then $\mathfrak{a}$ has no embedded prime ideals.
3. If A is absolutely flat, every primary ideal is maximal.
4. In the polynomial ring $\mathbf{Z}[t]$, the ideal $\mathfrak{m} = (2, t)$ is maximal and the ideal $\mathfrak{q} = (4, t)$ is $\mathfrak{m}$-primary, but is not a power of $\mathfrak{m}$.
5. In the polynomial ring $K[x, y, z]$ where K is a field and x, y, z are independent indeterminates, let $\mathfrak{p}_1 = (x, y)$, $\mathfrak{p}_2 = (x, z)$, $\mathfrak{m} = (x, y, z)$; $\mathfrak{p}_1$ and $\mathfrak{p}_2$ are prime, and $\mathfrak{m}$ is maximal. Let $\mathfrak{a} = \mathfrak{p}_1\mathfrak{p}_2$. Show that $\mathfrak{a} = \mathfrak{p}_1 \cap \mathfrak{p}_2 \cap \mathfrak{m}^2$ is a reduced primary decomposition of $\mathfrak{a}$. Which components are isolated and which are embedded?
6. Let X be an infinite compact Hausdorff space, $C(X)$ the ring of real-valued continuous functions on X (Chapter 1, Exercise 26). Is the zero ideal decomposable in this ring?
7. Let A be a ring and let $A[x]$ denote the ring of polynomials in one indeterminate over A. For each ideal $\mathfrak{a}$ of A, let $\mathfrak{a}[x]$ denote the set of all polynomials in $A[x]$ with coefficients in $\mathfrak{a}$.
   i) $\mathfrak{a}[x]$ is the extension of $\mathfrak{a}$ to $A[x]$.
   ii) If $\mathfrak{p}$ is a prime ideal in A, then $\mathfrak{p}[x]$ is a prime ideal in $A[x]$.
   iii) If $\mathfrak{q}$ is a $\mathfrak{p}$-primary ideal in A, then $\mathfrak{q}[x]$ is a $\mathfrak{p}[x]$-primary ideal in $A[x]$. [Use Chapter 1, Exercise 2.]
   iv) If $\mathfrak{a} = \bigcap_{i=1}^{n} \mathfrak{q}_i$ is a minimal primary decomposition in A, then $\mathfrak{a}[x] = \bigcap_{i=1}^{n} \mathfrak{q}_i[x]$ is a minimal primary decomposition in $A[x]$.
   v) If $\mathfrak{p}$ is a minimal prime ideal of $\mathfrak{a}$, then $\mathfrak{p}[x]$ is a minimal prime ideal of $\mathfrak{a}[x]$.
8. Let k be a field. Show that in the polynomial ring $k[x_1, \ldots, x_n]$ the ideals $\mathfrak{p}_i = (x_1, \ldots, x_i)$ $(1 \leqslant i \leqslant n)$ are prime and all their powers are primary. [Use Exercise 7.]
9. In a ring A, let $D(A)$ denote the set of prime ideals $\mathfrak{p}$ which satisfy the following condition: there exists $a \in A$ such that $\mathfrak{p}$ is minimal in the set of prime ideals containing $(0:a)$. Show that $x \in A$ is a zero divisor $\Leftrightarrow x \in \mathfrak{p}$ for some $\mathfrak{p} \in D(A)$.

   Let S be a multiplicatively closed subset of A, and identify Spec $(S^{-1}A)$ with its image in Spec (A) (Chapter 3, Exercise 21). Show that
   $$D(S^{-1}A) = D(A) \cap \text{Spec}\,(S^{-1}A).$$
   If the zero ideal has a primary decomposition, show that $D(A)$ is the set of associated prime ideals of 0.
10. For any prime ideal $\mathfrak{p}$ in a ring A, let $S_{\mathfrak{p}}(0)$ denote the kernel of the homomorphism $A \to A_{\mathfrak{p}}$. Prove that
    i) $S_{\mathfrak{p}}(0) \subseteq \mathfrak{p}$.
    ii) $r(S_{\mathfrak{p}}(0)) = \mathfrak{p} \Leftrightarrow \mathfrak{p}$ is a minimal prime ideal of A.
    iii) If $\mathfrak{p} \supseteq \mathfrak{p}'$, then $S_{\mathfrak{p}}(0) \subseteq S_{\mathfrak{p}'}(0)$.
    iv) $\bigcap_{\mathfrak{p} \in D(A)} S_{\mathfrak{p}}(0) = 0$, where $D(A)$ is defined in Exercise 9.

11. If $\mathfrak{p}$ is a minimal prime ideal of a ring A, show that $S_{\mathfrak{p}}(0)$ (Exercise 10) is the smallest $\mathfrak{p}$-primary ideal.

Let $\mathfrak{a}$ be the intersection of the ideals $S_{\mathfrak{p}}(0)$ as $\mathfrak{p}$ runs through the minimal prime ideals of A. Show that $\mathfrak{a}$ is contained in the nilradical of A.

Suppose that the zero ideal is decomposable. Prove that $\mathfrak{a} = 0$ if and only if every prime ideal of 0 is isolated.

12. Let A be a ring, S a multiplicatively closed subset of A. For any ideal $\mathfrak{a}$, let $S(\mathfrak{a})$ denote the contraction of $S^{-1}\mathfrak{a}$ in A. The ideal $S(\mathfrak{a})$ is called the *saturation* of $\mathfrak{a}$ with respect to S. Prove that

i) $S(\mathfrak{a}) \cap S(\mathfrak{b}) = S(\mathfrak{a} \cap \mathfrak{b})$

ii) $S(r(\mathfrak{a})) = r(S(\mathfrak{a}))$

iii) $S(\mathfrak{a}) = (1) \Leftrightarrow \mathfrak{a}$ meets S

iv) $S_1(S_2(\mathfrak{a})) = (S_1S_2)(\mathfrak{a})$.

If $\mathfrak{a}$ has a primary decomposition, prove that the set of ideals $S(\mathfrak{a})$ (where S runs through all multiplicatively closed subsets of A) is finite.

13. Let A be a ring and $\mathfrak{p}$ a prime ideal of A. The *nth symbolic power of* $\mathfrak{p}$ is defined to be the ideal (in the notation of Exercise 12)

$$\mathfrak{p}^{(n)} = S_{\mathfrak{p}}(\mathfrak{p}^n)$$

where $S_{\mathfrak{p}} = A - \mathfrak{p}$. Show that

i) $\mathfrak{p}^{(n)}$ is a $\mathfrak{p}$-primary ideal;

ii) if $\mathfrak{p}^n$ has a primary decomposition, then $\mathfrak{p}^{(n)}$ is its $\mathfrak{p}$-primary component;

iii) if $\mathfrak{p}^{(m)}\mathfrak{p}^{(n)}$ has a primary decomposition, then $\mathfrak{p}^{(m+n)}$ is its $\mathfrak{p}$-primary component;

iv) $\mathfrak{p}^{(n)} = \mathfrak{p}^n \Leftrightarrow \mathfrak{p}^{(n)}$ is $\mathfrak{p}$-primary.

14. Let $\mathfrak{a}$ be a decomposable ideal in a ring A and let $\mathfrak{p}$ be a maximal element of the set of ideals $(\mathfrak{a}:x)$, where $x \in A$ and $x \notin \mathfrak{a}$. Show that $\mathfrak{p}$ is a prime ideal belonging to $\mathfrak{a}$.

15. Let $\mathfrak{a}$ be a decomposable ideal in a ring A, let Σ be an isolated set of prime ideals belonging to $\mathfrak{a}$, and let $\mathfrak{q}_{\Sigma}$ be the intersection of the corresponding primary components. Let f be an element of A such that, for each prime ideal $\mathfrak{p}$ belonging to $\mathfrak{a}$, we have $f \in \mathfrak{p} \Leftrightarrow \mathfrak{p} \notin \Sigma$, and let S_f be the set of all powers of f. Show that $\mathfrak{q}_{\Sigma} = S_f(\mathfrak{a}) = (\mathfrak{a}:f^n)$ for all large n.

16. If A is a ring in which every ideal has a primary decomposition, show that every ring of fractions $S^{-1}A$ has the same property.

17. Let A be a ring with the following property.

(L1) For every ideal $\mathfrak{a} \neq (1)$ in A and every prime ideal $\mathfrak{p}$, there exists $x \notin \mathfrak{p}$ such that $S_{\mathfrak{p}}(\mathfrak{a}) = (\mathfrak{a}:x)$, where $S_{\mathfrak{p}} = A - \mathfrak{p}$.

Then every ideal in A is an intersection of (possibly infinitely many) primary ideals.

[Let $\mathfrak{a}$ be an ideal $\neq (1)$ in A, and let $\mathfrak{p}_1$ be a minimal element of the set of prime ideals containing $\mathfrak{a}$. Then $\mathfrak{q}_1 = S_{\mathfrak{p}_1}(\mathfrak{a})$ is $\mathfrak{p}_1$-primary (by Exercise 11), and $\mathfrak{q}_1 = (\mathfrak{a}:x)$ for some $x \notin \mathfrak{p}_1$. Show that $\mathfrak{a} = \mathfrak{q}_1 \cap (\mathfrak{a} + (x))$.

Now let $\mathfrak{a}_1$ be a maximal element of the set of ideals $\mathfrak{b} \supseteq \mathfrak{a}$ such that $\mathfrak{q}_1 \cap \mathfrak{b} = \mathfrak{a}$, and choose $\mathfrak{a}_1$ so that $x \in \mathfrak{a}_1$, and therefore $\mathfrak{a}_1 \not\subseteq \mathfrak{p}_1$. Repeat the construction starting with $\mathfrak{a}_1$, and so on. At the nth stage we have $\mathfrak{a} = \mathfrak{q}_1 \cap \cdots \cap \mathfrak{q}_n \cap \mathfrak{a}_n$ where the $\mathfrak{q}_i$ are primary ideals, $\mathfrak{a}_n$ is maximal among the ideals $\mathfrak{b}$ containing $\mathfrak{a}_{n-1} = \mathfrak{a}_n \cap \mathfrak{q}_n$ such that $\mathfrak{a} = \mathfrak{q}_1 \cap \cdots \cap \mathfrak{q}_n \cap \mathfrak{b}$, and $\mathfrak{a}_n \nsubseteq \mathfrak{p}_n$. If at any stage we have $\mathfrak{a}_n = (1)$, the process stops, and $\mathfrak{a}$ is a finite intersection of primary ideals. If not, continue by transfinite induction, observing that each $\mathfrak{a}_n$ strictly contains $\mathfrak{a}_{n-1}$.]

18. Consider the following condition on a ring A:
(L2) Given an ideal $\mathfrak{a}$ and a descending chain $S_1 \supseteq S_2 \supseteq \cdots \supseteq S_n \supseteq \cdots$ of multiplicatively closed subsets of A, there exists an integer n such that $S_n(\mathfrak{a}) = S_{n+1}(\mathfrak{a}) = \cdots$. Prove that the following are equivalent:
i) Every ideal in A has a primary decomposition;
ii) A satisfies (L1) and (L2).
[For i) ⇒ ii), use Exercises 12 and 15. For ii) ⇒ i) show, with the notation of the proof of Exercise 17, that if $S_n = S_{\mathfrak{p}_1} \cap \cdots \cap S_{\mathfrak{p}_n}$ then S_n meets $\mathfrak{a}_n$, hence $S_n(\mathfrak{a}_n) = (1)$, and therefore $S_n(\mathfrak{a}) = \mathfrak{q}_1 \cap \cdots \cap \mathfrak{q}_n$. Now use (L2) to show that the construction must terminate after a finite number of steps.]

19. Let A be a ring and $\mathfrak{p}$ a prime ideal of A. Show that every $\mathfrak{p}$-primary ideal contains $S_\mathfrak{p}(0)$, the kernel of the canonical homomorphism $A \to A_\mathfrak{p}$.

Suppose that A satisfies the following condition: for every prime ideal $\mathfrak{p}$, the intersection of all $\mathfrak{p}$-primary ideals of A is equal to $S_\mathfrak{p}(0)$. (Noetherian rings satisfy this condition: see Chapter 10.) Let $\mathfrak{p}_1, \ldots, \mathfrak{p}_n$ be distinct prime ideals, none of which is a minimal prime ideal of A. Then there exists an ideal $\mathfrak{a}$ in A whose associated prime ideals are $\mathfrak{p}_1, \ldots, \mathfrak{p}_n$.
[Proof by induction on n. The case $n = 1$ is trivial (take $\mathfrak{a} = \mathfrak{p}_1$). Suppose $n > 1$ and let $\mathfrak{p}_n$ be maximal in the set $\{\mathfrak{p}_1, \ldots, \mathfrak{p}_n\}$. By the inductive hypothesis there exists an ideal $\mathfrak{b}$ and a minimal primary decomposition $\mathfrak{b} = \mathfrak{q}_1 \cap \cdots \cap \mathfrak{q}_{n-1}$, where each $\mathfrak{q}_i$ is $\mathfrak{p}_i$-primary. If $\mathfrak{b} \subseteq S_{\mathfrak{p}_n}(0)$, let $\mathfrak{p}$ be a minimal prime ideal of A contained in $\mathfrak{p}_n$. Then $S_{\mathfrak{p}_n}(0) \subseteq S_\mathfrak{p}(0)$, hence $\mathfrak{b} \subseteq S_\mathfrak{p}(0)$. Taking radicals and using Exercise 10, we have $\mathfrak{p}_1 \cap \cdots \cap \mathfrak{p}_{n-1} \subseteq \mathfrak{p}$, hence some $\mathfrak{p}_i \subseteq \mathfrak{p}$, hence $\mathfrak{p}_i = \mathfrak{p}$ since $\mathfrak{p}$ is minimal. This is a contradiction since no $\mathfrak{p}_i$ is minimal. Hence $\mathfrak{b} \nsubseteq S_{\mathfrak{p}_n}(0)$ and therefore there exists a $\mathfrak{p}_n$-primary ideal $\mathfrak{q}_n$ such that $\mathfrak{b} \nsubseteq \mathfrak{q}_n$. Show that $\mathfrak{a} = \mathfrak{q}_1 \cap \cdots \cap \mathfrak{q}_n$ has the required properties.]

Primary decomposition of modules

Practically the whole of this chapter can be transposed to the context of modules over a ring A. The following exercises indicate how this is done.

20. Let M be a fixed A-module, N a submodule of M. The *radical* of N in M is defined to be

$$r_M(N) = \{x \in A : x^q M \subseteq N \text{ for some } q > 0\}.$$

Show that $r_M(N) = r(N:M) = r(\mathrm{Ann}\,(M/N))$. In particular, $r_M(N)$ is an *ideal*.

State and prove the formulas for r_M analogous to (1.13).

21. An element $x \in A$ defines an endomorphism ϕ_x of M, namely $m \mapsto xm$. The element x is said to be a *zero-divisor* (resp. *nilpotent*) *in* M if ϕ_x is not injective (resp. is nilpotent). A submodule Q of M is *primary in* M if $Q \neq M$ and every zero-divisor in M/Q is nilpotent.

Show that if Q is primary in M, then $(Q:M)$ is a primary ideal and hence $r_M(Q)$ is a prime ideal $\mathfrak{p}$. We say that Q is $\mathfrak{p}$-*primary* (in M).

Prove the analogues of (4.3) and (4.4).

22. A *primary decomposition of* N *in* M is a representation of N as an intersection

$$N = Q_1 \cap \cdots \cap Q_n$$

of primary submodules of M; it is a *minimal primary decomposition* if the ideals $\mathfrak{p}_i = r_M(Q_i)$ are all distinct and if none of the components Q_i can be omitted from the intersection, that is if $Q_i \not\supseteq \bigcap_{j \neq i} Q_j$ $(1 \leqslant i \leqslant n)$.

Prove the analogue of (4.5), that the prime ideals $\mathfrak{p}_i$ depend only on N (and M). They are called the *prime ideals belonging to* N *in* M. Show that they are also the prime ideals belonging to 0 in M/N.

23. State and prove the analogues of (4.6)–(4.11) inclusive. (There is no loss of generality in taking $N = 0$.)

5

Integral Dependence and Valuations

In classical algebraic geometry curves were frequently studied by projecting them onto a line and regarding the curve as a (ramified) covering of the line. This is quite analogous to the relationship between a number field and the rational field—or rather between their rings of integers—and the common algebraic feature is the notion of integral dependence. In this chapter we prove a number of results about integral dependence. In particular we prove the theorems of Cohen–Seidenberg (the "going-up" and "going-down" theorems) concerning prime ideals in an integral extension. In the exercises at the end we discuss the algebro-geometric situation and in particular the Normalization Lemma.

We also give a brief treatment of valuations.

INTEGRAL DEPENDENCE

Let B be a ring, A a subring of B (so that $1 \in A$). An element x of B is said to be *integral* over A if x is a root of a *monic* polynomial with coefficients in A, that is if x satisfies an equation of the form

$$x^n + a_1x^{n-1} + \cdots + a_n = 0 \tag{1}$$

where the a_i are elements of A. Clearly every element of A is integral over A.

Example 5.0. $A = \mathbf{Z}, B = \mathbf{Q}$. If a rational number $x = r/s$ is integral over $\mathbf{Z}$, where r, s have no common factor, we have from (1)

$$r^n + a_1r^{n-1}s + \cdots + a_ns^n = 0$$

the a_i being rational integers. Hence s divides r^n, hence $s = \pm 1$, hence $x \in \mathbf{Z}$.

Proposition 5.1. *The following are equivalent:*

i) *$x \in B$ is integral over A;*

ii) *$A[x]$ is a finitely generated A-module;*

iii) *$A[x]$ is contained in a subring C of B such that C is a finitely generated A-module;*

iv) *There exists a faithful $A[x]$-module M which is finitely generated as an A-module.*

Proof. i) $\Rightarrow$ ii). From (1) we have

$$x^{n+r} = -(a_1x^{n+r-1} + \cdots + a_nx^r)$$

for all $r \geq 0$; hence, by induction, all positive powers of x lie in the A-module generated by $1, x, \ldots, x^{n-1}$. Hence $A[x]$ is generated (as an A-module) by $1, x, \ldots, x^{n-1}$.

ii) $\Rightarrow$ iii). Take $C = A[x]$.

iii) $\Rightarrow$ iv). Take $M = C$, which is a faithful $A[x]$-module (since $yC = 0 \Rightarrow y \cdot 1 = 0$).

iv) $\Rightarrow$ i). This follows from (2.4): take ϕ to be multiplication by x, and $\mathfrak{a} = A$ (we have $xM \subseteq M$ since M is an $A[x]$-module); since M is faithful, we have $x^n + a_1x^{n-1} + \cdots + a_n = 0$ for suitable $a_i \in A$. ∎

Corollary 5.2. *Let x_i $(1 \leq i \leq n)$ be elements of B, each integral over A. Then the ring $A[x_1, \ldots, x_n]$ is a finitely-generated A-module.*

Proof. By induction on n. The case $n = 1$ is part of (5.1). Assume $n > 1$, let $A_r = A[x_1, \ldots, x_r]$; then by the inductive hypothesis A_{n-1} is a finitely generated A-module. $A_n = A_{n-1}[x_n]$ is a finitely generated A_{n-1}-module (by the case $n = 1$, since x_n is integral over A_{n-1}). Hence by (2.16) A_n is finitely generated as an A-module. ∎

Corollary 5.3. *The set C of elements of B which are integral over A is a subring of B containing A.*

Proof. If $x, y \in C$ then $A[x, y]$ is a finitely generated A-module by (5.2). Hence $x \pm y$ and xy are integral over A, by iii) of (5.1). ∎

The ring C in (5.3) is called the ***integral closure*** of A in B. If $C = A$, then A is said to be ***integrally closed in B***. If $C = B$, the ring B is said to be ***integral over A***.

Remark. Let $f: A \to B$ be a ring homomorphism, so that B is an A-algebra. Then f is said to be ***integral***, and B is said to be an ***integral*** A-algebra, if B is integral over its subring $f(A)$. In this terminology, the above results show that

$$\text{finite type} + \text{integral} = \text{finite}.$$

Corollary 5.4. *If $A \subseteq B \subseteq C$ are rings and if B is integral over A, and C is integral over B, then C is integral over A (transitivity of integral dependence).*

Proof. Let $x \in C$, then we have an equation

$$x^n + b_1x^{n-1} + \cdots + b_n = 0 \qquad (b_i \in B).$$

The ring $B' = A[b_1, \ldots, b_n]$ is a finitely generated A-module by (5.2), and $B'[x]$ is a finitely generated B'-module (since x is integral over B'). Hence $B'[x]$ is a finitely generated A-module by (2.16) and therefore x is integral over A by iii) of (5.1). ∎

Corollary 5.5. *Let $A \subseteq B$ be rings and let C be the integral closure of A in B. Then C is integrally closed in B.*

Proof. Let $x \in B$ be integral over C. By (5.4) x is integral over A, hence $x \in C$. ∎

The next proposition shows that integral dependence is preserved on passing to quotients and to rings of fractions:

Proposition 5.6. *Let $A \subseteq B$ be rings, B integral over A.*

i) If $\mathfrak{b}$ is an ideal of B and $\mathfrak{a} = \mathfrak{b}^c = A \cap \mathfrak{b}$, then $B/\mathfrak{b}$ is integral over $A/\mathfrak{a}$.

ii) If S is a multiplicatively closed subset of A, then $S^{-1}B$ is integral over $S^{-1}A$.

Proof. i) If $x \in B$ we have, say, $x^n + a_1x^{n-1} + \cdots + a_n = 0$, with $a_i \in A$. Reduce this equation mod. $\mathfrak{b}$.

ii) Let $x/s \in S^{-1}B(x \in B, s \in S)$. Then the equation above gives

$$(x/s)^n + (a_1/s)(x/s)^{n-1} + \cdots + a_n/s^n = 0$$

which shows that x/s is integral over $S^{-1}A$. ∎

THE GOING-UP THEOREM

Proposition 5.7. *Let $A \subseteq B$ be integral domains, B integral over A. Then B is a field if and only if A is a field.*

Proof. Suppose A is a field; let $y \in B$, $y \neq 0$. Let

$$y^n + a_1y^{n-1} + \cdots + a_n = 0 \qquad (a_i \in A)$$

be an equation of integral dependence for y of smallest possible degree. Since B is an integral domain we have $a_n \neq 0$, hence $y^{-1} = -a_n^{-1}(y^{n-1} + a_1y^{n-2} + \cdots + a_{n-1}) \in B$. Hence B is a field.

Conversely, suppose B is a field; let $x \in A$, $x \neq 0$. Then $x^{-1} \in B$, hence is integral over A, so that we have an equation

$$x^{-m} + a_1'x^{-m+1} + \cdots + a_m' = 0 \qquad (a_i' \in A).$$

It follows that $x^{-1} = -(a_1' + a_2'x + \cdots + a_m'x^{m-1}) \in A$, hence A is a field. ∎

Corollary 5.8. *Let $A \subseteq B$ be rings, B integral over A; let $\mathfrak{q}$ be a prime ideal of B and let $\mathfrak{p} = \mathfrak{q}^c = \mathfrak{q} \cap A$. Then $\mathfrak{q}$ is maximal if and only if $\mathfrak{p}$ is maximal.*

Proof. By (5.6), $B/\mathfrak{q}$ is integral over $A/\mathfrak{p}$, and both these rings are integral domains. Now use (5.7). ∎

Corollary 5.9. *Let $A \subseteq B$ be rings, B integral over A; let $\mathfrak{q}, \mathfrak{q}'$ be prime ideals of B such that $\mathfrak{q} \subseteq \mathfrak{q}'$ and $\mathfrak{q}^c = \mathfrak{q}'^c = \mathfrak{p}$ say. Then $\mathfrak{q} = \mathfrak{q}'$.*

Proof. By (5.6), $B_\mathfrak{p}$ is integral over $A_\mathfrak{p}$. Let $\mathfrak{m}$ be the extension of $\mathfrak{p}$ in $A_\mathfrak{p}$ and let $\mathfrak{n}, \mathfrak{n}'$ be the extensions of $\mathfrak{q}, \mathfrak{q}'$ respectively in $B_\mathfrak{p}$. Then $\mathfrak{m}$ is the maximal ideal of $A_\mathfrak{p}$; $\mathfrak{n} \subseteq \mathfrak{n}'$, and $\mathfrak{n}^c = \mathfrak{n}'^c = \mathfrak{m}$. By (5.8) it follows that $\mathfrak{n}, \mathfrak{n}'$ are maximal, hence $\mathfrak{n} = \mathfrak{n}'$, hence by (3.11)(iv) $\mathfrak{q} = \mathfrak{q}'$. ∎

Theorem 5.10. *Let $A \subseteq B$ be rings, B integral over A, and let $\mathfrak{p}$ be a prime ideal of A. Then there exists a prime ideal $\mathfrak{q}$ of B such that $\mathfrak{q} \cap A = \mathfrak{p}$.*

Proof. By (5.6), $B_\mathfrak{p}$ is integral over $A_\mathfrak{p}$, and the diagram

$$\begin{array}{ccc} A & \rightarrow & B \\ {\scriptstyle\alpha}\downarrow & & \downarrow{\scriptstyle\beta} \\ A_\mathfrak{p} & \rightarrow & B_\mathfrak{p} \end{array}$$

(in which the horizontal arrows are injections) is commutative. Let $\mathfrak{n}$ be a maximal ideal of $B_\mathfrak{p}$; then $\mathfrak{m} = \mathfrak{n} \cap A_\mathfrak{p}$ is maximal by (5.8), hence is the unique maximal ideal of the local ring $A_\mathfrak{p}$. If $\mathfrak{q} = \beta^{-1}(\mathfrak{n})$, then $\mathfrak{q}$ is prime and we have $\mathfrak{q} \cap A = \alpha^{-1}(\mathfrak{m}) = \mathfrak{p}$. ∎

Theorem 5.11. ("Going-up theorem"). *Let $A \subseteq B$ be rings, B integral over A; let $\mathfrak{p}_1 \subseteq \cdots \subseteq \mathfrak{p}_n$ be a chain of prime ideals of A and $\mathfrak{q} \subseteq \cdots \subseteq \mathfrak{q}_m$ $(m < n)$ a chain of prime ideals of B such that $\mathfrak{q}_i \cap A = \mathfrak{p}_i$ $(1 \leqslant i \leqslant m)$. Then the chain $\mathfrak{q}_1 \subseteq \cdots \subseteq \mathfrak{q}_m$ can be extended to a chain $\mathfrak{q}_1 \subseteq \cdots \subseteq \mathfrak{q}_n$ such that $\mathfrak{q}_i \cap A = \mathfrak{p}_i$ for $1 \leqslant i \leqslant n$.*

Proof. By induction we reduce immediately to the case $m = 1$, $n = 2$. Let $\bar{A} = A/\mathfrak{p}_1$, $\bar{B} = B/\mathfrak{q}_1$; then $\bar{A} \subseteq \bar{B}$, and $\bar{B}$ is integral over $\bar{A}$ by (5.6). Hence, by (5.10), there exists a prime ideal $\bar{\mathfrak{q}}_2$ of $\bar{B}$ such that $\bar{\mathfrak{q}}_2 \cap \bar{A} = \bar{\mathfrak{p}}_2$, the image of $\mathfrak{p}_2$ in $\bar{A}$. Lift back $\bar{\mathfrak{q}}_2$ to B and we have a prime ideal $\mathfrak{q}_2$ with the required properties. ∎

INTEGRALLY CLOSED INTEGRAL DOMAINS. THE GOING-DOWN THEOREM

Proposition (5.6)(ii) can be sharpened:

Proposition 5.12. *Let $A \subseteq B$ be rings, C the integral closure of A in B. Let S be a multiplicatively closed subset of A. Then $S^{-1}C$ is the integral closure of $S^{-1}A$ in $S^{-1}B$.*

Proof. By (5.6), $S^{-1}C$ is integral over $S^{-1}A$. Conversely, if $b/s \in S^{-1}B$ is integral over $S^{-1}A$, then we have an equation of the form

$$(b/s)^n + (a_1/s_1)(b/s)^{n-1} + \cdots + a_n/s_n = 0$$

where $a_i \in A$, $s_i \in S$ $(1 \leqslant i \leqslant n)$. Let $t = s_1 \cdots s_n$ and multiply this equation by $(st)^n$ throughout. Then it becomes an equation of integral dependence for bt over A. Hence $bt \in C$ and therefore $b/s = bt/st \in S^{-1}C$. ∎

An integral domain is said to be *integrally closed* (without qualification) if it is integrally closed in its field of fractions. For example, $\mathbf{Z}$ is integrally closed (see (5.0)). The same argument shows that any unique factorization domain is integrally closed. In particular, a polynomial ring $k[x_1, \ldots, x_n]$ over a field is integrally closed.

Integral closure is a local property:

Proposition 5.13. *Let A be an integral domain. Then the following are equivalent:*

i) *A is integrally closed;*

ii) *$A_\mathfrak{p}$ is integrally closed, for each prime ideal $\mathfrak{p}$;*

iii) *$A_\mathfrak{m}$ is integrally closed, for each maximal ideal $\mathfrak{m}$.*

Proof. Let K be the field of fractions of A, let C be the integral closure of A in K, and let $f: A \to C$ be the identity mapping of A into C. Then A is integrally closed $\Leftrightarrow f$ is surjective, and by (5.12) $A_\mathfrak{p}$ (resp. $A_\mathfrak{m}$) is integrally closed $\Leftrightarrow f_\mathfrak{p}$ (resp. $f_\mathfrak{m}$) is surjective. Now use (3.9). ∎

Let $A \subseteq B$ be rings and let $\mathfrak{a}$ be an ideal of A. An element of B is said to be *integral over* $\mathfrak{a}$ if it satisfies an equation of integral dependence over A in which all the coefficients lie in $\mathfrak{a}$. The *integral closure* of $\mathfrak{a}$ in B is the set of all elements of B which are integral over $\mathfrak{a}$.

Lemma 5.14. *Let C be the integral closure of A in B and let $\mathfrak{a}^e$ denote the extension of $\mathfrak{a}$ in C. Then the integral closure of $\mathfrak{a}$ in B is the radical of $\mathfrak{a}^e$ (and is therefore closed under addition and multiplication).*

Proof. If $x \in B$ is integral over $\mathfrak{a}$, we have an equation of the form

$$x^n + a_1x^{n-1} + \cdots + a_n = 0$$

with $a_1, \ldots, a_n$ in $\mathfrak{a}$. Hence $x \in C$ and $x^n \in \mathfrak{a}^e$, that is $x \in r(\mathfrak{a}^e)$. Conversely, if $x \in r(\mathfrak{a}^e)$ then $x^n = \sum a_i x_i$ for some $n > 0$, where the a_i are elements of $\mathfrak{a}$ and the x_i are elements of C. Since each x_i is integral over A it follows from (5.2) that $M = A[x_1, \ldots, x_n]$ is a finitely generated A-module, and we have $x^nM \subseteq \mathfrak{a}M$. Hence by (2.4) (taking ϕ there to be multiplication by x^n) we see that x^n is integral over $\mathfrak{a}$, hence x is integral over $\mathfrak{a}$. ∎

Proposition 5.15. *Let $A \subseteq B$ be integral domains, A integrally closed, and let $x \in B$ be integral over an ideal $\mathfrak{a}$ of A. Then x is algebraic over the field of fractions K of A, and if its minimal polynomial over K is $t^n + a_1t^{n-1} + \cdots + a_n$, then $a_1, \ldots, a_n$ lie in $r(\mathfrak{a})$.*

Proof. Clearly x is algebraic over K. Let L be an extension field of K which contains all the conjugates $x_1, \ldots, x_n$ of x. Each x_i satisfies the same equation of integral dependence as x does, hence each x_i is integral over $\mathfrak{a}$. The coefficients of the minimal polynomial of x over K are polynomials in the x_i, hence by (5.14) are integral over $\mathfrak{a}$. Since A is integrally closed, they must lie in $r(\mathfrak{a})$, by (5.14) again. ∎

Theorem 5.16. ("Going-down theorem"). *Let $A \subseteq B$ be integral domains, A integrally closed, B integral over A. Let $\mathfrak{p}_1 \supseteq \cdots \supseteq \mathfrak{p}_n$ be a chain of prime ideals of A, and let $\mathfrak{q}_1 \supseteq \cdots \supseteq \mathfrak{q}_m$ $(m < n)$ be a chain of prime ideals of B such that $\mathfrak{q}_i \cap A = \mathfrak{p}_i$ $(1 \leqslant i \leqslant m)$. Then the chain $\mathfrak{q}_1 \supseteq \cdots \supseteq \mathfrak{q}_m$ can be extended to a chain $\mathfrak{q}_1 \supseteq \cdots \supseteq \mathfrak{q}_n$ such that $\mathfrak{q}_i \cap A = \mathfrak{p}_i$ $(1 \leqslant i \leqslant n)$.*

Proof. As in (5.11) we reduce immediately to the case $m = 1, n = 2$. Then we have to show that $\mathfrak{p}_2$ is the contraction of a prime ideal in the ring $B_{\mathfrak{q}_1}$, or equivalently (3.16) that $B_{\mathfrak{q}_1}\mathfrak{p}_2 \cap A = \mathfrak{p}_2$.

Every $x \in B_{\mathfrak{q}_1}\mathfrak{p}_2$ is of the form y/s, where $y \in B\mathfrak{p}_2$ and $s \in B - \mathfrak{q}_1$. By (5.14), y is integral over $\mathfrak{p}_2$, and hence by (5.15) its minimal equation over K, the field of fractions of A, is of the form

$$y^r + u_1 y^{r-1} + \cdots + u_r = 0 \tag{1}$$

with $u_1, \ldots, u_r$ in $\mathfrak{p}_2$.

Now suppose that $x \in B_{\mathfrak{q}_1}\mathfrak{p}_2 \cap A$. Then $s = yx^{-1}$ with $x^{-1} \in K$, so that the minimal equation for s over K is obtained by dividing (1) by x^r, and is therefore, say,

$$s^r + v_1 s^{r-1} + \cdots + v_r = 0 \tag{2}$$

where $v_i = u_i/x^i$. Consequently

$$x^i v_i = u_i \in \mathfrak{p}_2 \qquad (1 \leqslant i \leqslant r). \tag{3}$$

But s is integral over A, hence by (5.15) (with $\mathfrak{a} = (1)$) each v_i is in A. Suppose $x \notin \mathfrak{p}_2$. Then (3) shows that each $v_i \in \mathfrak{p}_2$, hence (2) shows that $s^r \in B\mathfrak{p}_2 \subseteq B\mathfrak{p}_1 \subseteq \mathfrak{q}_1$, and therefore $s \in \mathfrak{q}_1$, which is a contradiction. Hence $x \in \mathfrak{p}_2$ and therefore $B_{\mathfrak{q}_1}\mathfrak{p}_2 \cap A = \mathfrak{p}_2$ as required. ■

The proof of the next proposition assumes some standard facts from field theory.

Proposition 5.17. *Let A be an integrally closed domain, K its field of fractions, L a finite separable algebraic extension of K, B the integral closure of A in L. Then there exists a basis $v_1, \ldots, v_n$ of L over K such that $B \subseteq \sum_{j=1}^{n} Av_j$.*

Proof. If v is any element of L, then v is algebraic over K and therefore satisfies an equation of the form

$$a_0 v^r + a_1 v^{r-1} + \cdots + a_n = 0 \; (a_i \in A).$$

Multiplying this equation by a_0^{r-1}, we see that $a_0 v = u$ is integral over A, and hence is in B. Thus, given any basis of L over K we may multiply the basis elements by suitable elements of A to get a basis $u_1, \ldots, u_n$ such that each $u_i \in B$.

Let T denote trace (from L to K). Since L/K is separable, the bilinear form $(x, y) \mapsto T(xy)$ on L (considered as a vector space over K) is non-degenerate, and hence we have a dual basis $v_1, \ldots, v_n$ of L over K, defined by $T(u_i v_j) = \delta_{ij}$. Let $x \in B$, say $x = \sum_j x_j v_j$ $(x_j \in K)$. We have $xu_i \in B$ (since $u_i \in B$) and therefore $T(xu_i) \in A$ by (5.15) (for the trace of an element is a multiple of one of the coefficients in the minimal polynomial). But $T(xu_i) = \sum_j T(x_j u_i v_j) = \sum_j x_j T(u_i v_j) = \sum_j x_j \delta_{ij} = x_i$, hence $x_i \in A$. Consequently $B \subseteq \sum_j Av_j$. ■

VALUATION RINGS

Let B be an integral domain, K its field of fractions. B is a *valuation ring* of K if, for each $x \neq 0$, either $x \in B$ or $x^{-1} \in B$ (or both).

Proposition 5.18. i) *B is a local ring.*

ii) *If B' is a ring such that $B \subseteq B' \subseteq K$, then B' is a valuation ring of K.*

iii) *B is integrally closed (in K).*

Proof. i) Let $\mathfrak{m}$ be the set of non-units of B, so that $x \in \mathfrak{m} \Leftrightarrow$ either $x = 0$ or $x^{-1} \notin B$. If $a \in B$ and $x \in \mathfrak{m}$ we have $ax \in \mathfrak{m}$, for otherwise $(ax)^{-1} \in B$ and therefore $x^{-1} = a\cdot(ax)^{-1} \in B$. Next let x, y be non-zero elements of $\mathfrak{m}$. Then either $xy^{-1} \in B$ or $x^{-1}y \in B$. If $xy^{-1} \in B$ then $x + y = (1 + xy^{-1})y \in B\mathfrak{m} \subseteq \mathfrak{m}$, and similarly if $x^{-1}y \in B$. Hence $\mathfrak{m}$ is an ideal and therefore B is a local ring by (1.6).

ii) Clear from the definitions.

iii) Let $x \in K$ be integral over B. Then we have, say,

$$x^n + b_1x^{n-1} + \cdots + b_n = 0$$

with the $b_i \in B$. If $x \in B$ there is nothing to prove. If not, then $x^{-1} \in B$, hence $x = -(b_1 + b_2x^{-1} + \cdots + b_nx^{1-n}) \in B$.

Let K be a field, Ω an algebraically closed field. Let Σ be the set of all pairs (A, f), where A is a subring of K and f is a homomorphism of A into Ω. We partially order the set Σ as follows:

$$(A, f) \leqslant (A', f') \Leftrightarrow A \subseteq A' \text{ and } f'|A = f.$$

The conditions of Zorn's lemma are clearly satisfied and therefore the set Σ has at least one maximal element.

Let (B, g) be a maximal element of Σ. We want to prove that B is a valuation ring of K. The first step in the proof is

Lemma 5.19. *B is a local ring and $\mathfrak{m} = \operatorname{Ker}(g)$ is its maximal ideal.*

Proof. Since $g(B)$ is a subring of a field and therefore an integral domain, the ideal $\mathfrak{m} = \operatorname{Ker}(g)$ is prime. We can extend g to a homomorphism $\bar{g}: B_\mathfrak{m} \to \Omega$ by putting $\bar{g}(b/s) = g(b)/g(s)$ for all $b \in B$ and all $s \in B - \mathfrak{m}$, since $g(s)$ will not be zero. Since the pair (B, g) is maximal it follows that $B = B_\mathfrak{m}$, hence B is a local ring and $\mathfrak{m}$ is its maximal ideal. ∎

Lemma 5.20. *Let x be a non-zero element of K. Let $B[x]$ be the subring of K generated by x over B, and let $\mathfrak{m}[x]$ be the extension of $\mathfrak{m}$ in $B[x]$. Then either $\mathfrak{m}[x] \neq B[x]$ or $\mathfrak{m}[x^{-1}] \neq B[x^{-1}]$.*

Proof. Suppose that $\mathfrak{m}[x] = B[x]$ and $\mathfrak{m}[x^{-1}] = B[x^{-1}]$. Then we shall have equations

$$u_0 + u_1x + \cdots + u_mx^m = 1 \qquad (u_i \in \mathfrak{m}) \tag{1}$$

$$v_0 + v_1x^{-1} + \cdots + v_nx^{-n} = 1 \qquad (v_j \in \mathfrak{m}) \tag{2}$$

in which we may assume that the degrees m, n are as small as possible. Suppose that $m \geqslant n$, and multiply (2) through by x^n:

$$(1 - v_0)x^n = v_1x^{n-1} + \cdots + v_n. \tag{3}$$

Since $v_0 \in \mathfrak{m}$, it follows from (5.19) that $1 - v_0$ is a unit in B, and (3) may therefore be written in the form

$$x^n = w_1 x^{n-1} + \cdots + w_n \qquad (w_i \in \mathfrak{m}).$$

Hence we can replace x^m in (1) by $w_1 x^{m-1} + \cdots + w_n x^{m-n}$, and this contradicts the minimality of the exponent m. ∎

Theorem 5.21. Let (B, g) be a maximal element of Σ. Then B is a valuation ring of the field K.

Proof. We have to show that if $x \neq 0$ is an element of K, then either $x \in B$ or $x^{-1} \in B$. By (5.20) we may as well assume that $\mathfrak{m}[x]$ is not the unit ideal of the ring $B' = B[x]$. Then $\mathfrak{m}[x]$ is contained in a maximal ideal $\mathfrak{m}'$ of B', and we have $\mathfrak{m}' \cap B = \mathfrak{m}$ (because $\mathfrak{m}' \cap B$ is a proper ideal of B and contains $\mathfrak{m}$). Hence the embedding of B in B' induces an embedding of the field $k = B/\mathfrak{m}$ in the field $k' = B'/\mathfrak{m}'$; also $k' = k[\bar{x}]$ where $\bar{x}$ is the image of x in k', hence $\bar{x}$ is algebraic over k, and therefore k' is a finite algebraic extension of k.

Now the homomorphism g induces an embedding $\bar{g}$ of k in Ω, since by (5.19) $\mathfrak{m}$ is the kernel of g. Since Ω is algebraically closed, $\bar{g}$ can be extended to an embedding $\bar{g}'$ of k' into Ω. Composing $\bar{g}'$ with the natural homomorphism $B' \to k'$, we have, say, $g' : B' \to \Omega$ which extends g. Since the pair (B, g) is maximal, it follows that $B' = B$ and therefore $x \in B$. ∎

Corollary 5.22. Let A be a subring of a field K. Then the integral closure $\bar{A}$ of A in K is the intersection of all the valuation rings of K which contain A.

Proof. Let B be a valuation ring of K such that $A \subseteq B$. Since B is integrally closed, by (5.18) iii), it follows that $\bar{A} \subseteq B$.

Conversely, let $x \notin \bar{A}$. Then x is not in the ring $A' = A[x^{-1}]$. Hence x^{-1} is a non-unit in A' and is therefore contained in a maximal ideal $\mathfrak{m}'$ of A'. Let Ω be an algebraic closure of the field $k' = A'/\mathfrak{m}'$. Then the restriction to A of the natural homomorphism $A' \to k'$ defines a homomorphism of A into Ω. By (5.21) this can be extended to some valuation ring $B \supseteq A$. Since x^{-1} maps to zero, it follows that $x \notin B$. ∎

Proposition 5.23. Let $A \subseteq B$ be integral domains, B finitely generated over A. Let v be a non-zero element of B. Then there exists $u \neq 0$ in A with the following property: any homomorphism f of A into an algebraically closed field Ω such that $f(u) \neq 0$ can be extended to a homomorphism g of B into Ω such that $g(v) \neq 0$.

Proof. By induction on the number of generators of B over A we reduce immediately to the case where B is generated over A by a single element x.

i) Suppose x is transcendental over A, i.e., that no non-zero polynomial with coefficients in A has x as a root. Let $v = a_0 x^n + a_1 x^{n-1} + \cdots + a_n$, and take $u = a_0$. Then if $f : A \to \Omega$ is such that $f(u) \neq 0$, there exists $\xi \in \Omega$ such that $f(a_0)\xi^n + f(a_1)\xi^{n-1} + \cdots + f(a_n) \neq 0$, because Ω is infinite. Define $g : B \to \Omega$ extending f by putting $g(x) = \xi$. Then $g(v) \neq 0$, as required.

ii) Now suppose x is algebraic over A (i.e. over the field of fractions of A). Then so is v^{-1}, because v is a polynomial in x. Hence we have equations of the form

$$a_0 x^m + a_1 x^{m-1} + \cdots + a_m = 0 \qquad (a_i \in A) \tag{1}$$

$$a_0' v^{-n} + a_1' v^{1-n} + \cdots + a_n' = 0 \qquad (a_j' \in A). \tag{2}$$

Let $u = a_0 a'_0$, and let $f: A \to \Omega$ be such that $f(u) \neq 0$. Then f can be extended, first to a homomorphism $f_1: A[u^{-1}] \to \Omega$ (with $f_1(u^{-1}) = f(u)^{-1}$), and then by (5.21) to a homomorphism $h: C \to \Omega$, where C is a valuation ring containing $A[u^{-1}]$. From (1), x is integral over $A[u^{-1}]$, hence by (5.22) $x \in C$, so that C contains B, and in particular $v \in C$. On the other hand, from (2), v^{-1} is integral over $A[u^{-1}]$, and therefore by (5.22) again is in C. Therefore v is a unit in C, and hence $h(v) \neq 0$. Now take g to be the restriction of h to B. ∎

Corollary 5.24. *Let k be a field and B a finitely generated k-algebra. If B is a field then it is a finite algebraic extension of k.*

Proof. Take $A = k$, $v = 1$ and Ω = algebraic closure of k. ∎

(5.24) is one form of Hilbert's Nullstellensatz. For another proof, see (7.9).

EXERCISES

1. Let $f: A \to B$ be an integral homomorphism of rings. Show that $f^*: \text{Spec}(B) \to \text{Spec}(A)$ is a *closed* mapping, i.e. that it maps closed sets to closed sets. (This is a geometrical equivalent of (5.10).)
2. Let A be a subring of a ring B such that B is integral over A, and let $f: A \to \Omega$ be a homomorphism of A into an algebraically closed field Ω. Show that f can be extended to a homomorphism of B into Ω. [Use (5.10).]
3. Let $f: B \to B'$ be a homomorphism of A-algebras, and let C be an A-algebra. If f is integral, prove that $f \otimes 1: B \otimes_A C \to B' \otimes_A C$ is integral. (This includes (5.6) ii) as a special case.)
4. Let A be a subring of a ring B such that B is integral over A. Let $\mathfrak{n}$ be a maximal ideal of B and let $\mathfrak{m} = \mathfrak{n} \cap A$ be the corresponding maximal ideal of A. Is $B_\mathfrak{n}$ necessarily integral over $A_\mathfrak{m}$?
[Consider the subring $k[x^2 - 1]$ of $k[x]$, where k is a field, and let $\mathfrak{n} = (x - 1)$. Can the element $1/(x + 1)$ be integral?]
5. Let $A \subseteq B$ be rings, B integral over A.
i) If $x \in A$ is a unit in B then it is a unit in A.
ii) The Jacobson radical of A is the contraction of the Jacobson radical of B.
6. Let $B_1, \ldots, B_n$ be integral A-algebras. Show that $\prod_{i=1}^n B_i$ is an integral A-algebra.
7. Let A be a subring of a ring B, such that the set $B - A$ is closed under multiplication. Show that A is integrally closed in B.
8. i) Let A be a subring of an integral domain B, and let C be the integral closure of A in B. Let f, g be monic polynomials in $B[x]$ such that $fg \in C[x]$. Then f, g are in $C[x]$. [Take a field containing B in which the polynomials f, g split into linear factors: say $f = \prod (x - \xi_i)$, $g = \prod (x - \eta_j)$. Each ξ_i and each η_j is a root of fg, hence is integral over C. Hence the coefficients of f and g are integral over C.]
ii) Prove the same result without assuming that B (or A) is an integral domain.

9. Let A be a subring of a ring B and let C be the integral closure of A in B. Prove that $C[x]$ is the integral closure of $A[x]$ in $B[x]$. [If $f \in B[x]$ is integral over $A[x]$, then

$$f^m + g_1 f^{m-1} + \cdots + g_m = 0 \qquad (g_i \in A[x]).$$

Let r be an integer larger than m and the degrees of $g_1, \ldots, g_m$, and let $f_1 = f - x^r$, so that

$$(f_1 + x^r)^m + g_1(f + x^r)^{m-1} + \cdots + g_m = 0$$

or say

$$f_1^m + h_1 f_1^{m-1} + \cdots + h_m = 0,$$

where $h_m = (x^r)^m + g_1(x^r)^{m-1} + \cdots + g_m \in A[x]$. Now apply Exercise 8 to the polynomials $-f_1$ and $f_1^{m-1} + h_1 f_1^{m-2} + \cdots + h_{m-1}$.]

10. A ring homomorphism $f: A \to B$ is said to have the *going-up property* (resp. the *going-down property*) if the conclusion of the going-up theorem (5.11) (resp. the going-down theorem (5.16)) holds for B and its subring $f(A)$.

Let $f^*: \operatorname{Spec}(B) \to \operatorname{Spec}(A)$ be the mapping associated with f.

i) Consider the following three statements:
   (a) f^* is a closed mapping.
   (b) f has the going-up property.
   (c) Let $\mathfrak{q}$ be any prime ideal of B and let $\mathfrak{p} = \mathfrak{q}^c$. Then $f^*: \operatorname{Spec}(B/\mathfrak{q}) \to \operatorname{Spec}(A/\mathfrak{p})$ is surjective.

   Prove that (a) $\Rightarrow$ (b) $\Leftrightarrow$ (c). (See also Chapter 6, Exercise 11.)

ii) Consider the following three statements:
   (a′) f^* is an open mapping.
   (b′) f has the going-down property.
   (c′) For any prime ideal $\mathfrak{q}$ of B, if $\mathfrak{p} = \mathfrak{q}^c$, then $f^*: \operatorname{Spec}(B_\mathfrak{q}) \to \operatorname{Spec}(A_\mathfrak{p})$ is surjective.

   Prove that (a′) $\Rightarrow$ (b′) $\Leftrightarrow$ (c′). (See also Chapter 7, Exercise 23.)

[To prove that (a′) $\Rightarrow$ (c′), observe that $B_\mathfrak{q}$ is the direct limit of the rings B_t where $t \in B - \mathfrak{q}$; hence, by Chapter 3, Exercise 26, we have $f^*(\operatorname{Spec}(B_\mathfrak{q})) = \bigcap_t f^*(\operatorname{Spec}(B_t)) = \bigcap_t f^*(Y_t)$. Since Y_t is an open neighborhood of $\mathfrak{q}$ in Y, and since f^* is open, it follows that $f^*(Y_t)$ is an open neighborhood of $\mathfrak{p}$ in X and therefore contains $\operatorname{Spec}(A_\mathfrak{p})$.]

11. Let $f: A \to B$ be a flat homomorphism of rings. Then f has the going-down property. [Chapter 3, Exercise 18.]

12. Let G be a finite group of automorphisms of a ring A, and let A^G denote the subring of G-invariants, that is of all $x \in A$ such that $\sigma(x) = x$ for all $\sigma \in G$. Prove that A is integral over A^G. [If $x \in A$, observe that x is a root of the polynomial $\prod_{\sigma \in G}(t - \sigma(x))$.]

Let S be a multiplicatively closed subset of A such that $\sigma(S) \subseteq S$ for all $\sigma \in G$, and let $S^G = S \cap A^G$. Show that the action of G on A extends to an action on $S^{-1}A$, and that $(S^G)^{-1}A^G \cong (S^{-1}A)^G$.

13. In the situation of Exercise 12, let $\mathfrak{p}$ be a prime ideal of A^G, and let P be the set of prime ideals of A whose contraction is $\mathfrak{p}$. Show that G acts transitively on P. In particular, P is *finite*.

[Let $\mathfrak{p}_1\mathfrak{p}_2 \in P$ and let $x \in \mathfrak{p}_1$. Then $\Pi_\sigma\, \sigma(x) \in \mathfrak{p}_1 \cap A^G = \mathfrak{p} \subseteq \mathfrak{p}_2$, hence $\sigma(x) \in \mathfrak{p}_2$ for some $\sigma \in G$. Deduce that $\mathfrak{p}_1$ is contained in $\bigcup_{\sigma \in G} \sigma(\mathfrak{p}_2)$, and then apply (1.11) and (5.9).]

14. Let A be an integrally closed domain, K its field of fractions and L a finite normal separable extension of K. Let G be the Galois group of L over K and let B be the integral closure of A in L. Show that $\sigma(B) = B$ for all $\sigma \in G$, and that $A = B^G$.

15. Let A, K be as in Exercise 14, let L be any finite extension field of K, and let B be the integral closure of A in L. Show that, if $\mathfrak{p}$ is any prime ideal of A, then the set of prime ideals $\mathfrak{q}$ of B which contract to $\mathfrak{p}$ is finite (in other words, that Spec $(B) \to$ Spec (A) has finite fibers).
[Reduce to the two cases (a) L separable over K and (b) L purely inseparable over K. In case (a), embed L in a finite normal separable extension of K, and use Exercises 13 and 14. In case (b), if $\mathfrak{q}$ is a prime ideal of B such that $\mathfrak{q} \cap A = \mathfrak{p}$, show that $\mathfrak{q}$ is the set of all $x \in B$ such that $x^{p^m} \in \mathfrak{p}$ for some $m \geq 0$, where p is the characteristic of K, and hence that Spec $(B) \to$ Spec (A) is bijective in this case.]

Noether's normalization lemma

16. Let k be a field and let $A \neq 0$ be a finitely generated k-algebra. Then there exist elements $y_1, \ldots, y_r \in A$ which are algebraically independent over k and such that A is integral over $k[y_1, \ldots, y_r]$.
We shall assume that k is *infinite*. (The result is still true if k is finite, but a different proof is needed.) Let $x_1, \ldots, x_n$ generate A as a k-algebra. We can renumber the x_i so that $x_1, \ldots, x_r$ are algebraically independent over k and each of $x_{r+1}, \ldots, x_n$ is algebraic over $k[x_1, \ldots, x_r]$. Now proceed by induction on n. If $n = r$ there is nothing to do, so suppose $n > r$ and the result true for $n - 1$ generators. The generator x_n is algebraic over $k[x_1, \ldots, x_{n-1}]$, hence there exists a polynomial $f \neq 0$ in n variables such that $f(x_1, \ldots, x_{n-1}, x_n) = 0$. Let F be the homogeneous part of highest degree in f. Since k is infinite, there exist $\lambda_1, \ldots, \lambda_{n-1} \in k$ such that $F(\lambda_1, \ldots, \lambda_{n-1}, 1) \neq 0$. Put $x_i' = x_i - \lambda_i x_n$ $(1 \leq i \leq n - 1)$. Show that x_n is integral over the ring $A' = k[x_1', \ldots, x_{n-1}']$, and hence that A is integral over A'. Then apply the inductive hypothesis to A' to complete the proof.
From the proof it follows that $y_1, \ldots, y_r$ may be chosen to be linear combinations of $x_1, \ldots, x_n$. This has the following geometrical interpretation: if k is algebraically closed and X is an affine algebraic variety in k^n with coordinate ring $A \neq 0$, then there exists a linear subspace L of dimension r in k^n and a linear mapping of k^n onto L which maps X *onto* L. [Use Exercise 2.]

Nullstellensatz (weak form).

17. Let X be an affine algebraic variety in k^n, where k is an algebraically closed field, and let $I(X)$ be the ideal of X in the polynomial ring $k[t_1, \ldots, t_n]$ (Chapter 1, Exercise 27). If $I(X) \neq (1)$ then X is not empty. [Let $A = k[t_1, \ldots, t_n]/I(X)$ be the coordinate ring of X. Then $A \neq 0$, hence by Exercise 16 there exists a linear subspace L of dimension ≥ 0 in k^n and a mapping of X *onto* L. Hence $X \neq \varnothing$.]

Deduce that every maximal ideal in the ring $k[t_1, \ldots, t_n]$ is of the form $(t_1 - a_1, \ldots, t_n - a_n)$ where $a_i \in k$.

18. Let k be a field and let B be a finitely generated k-algebra. Suppose that B is a field. Then B is a finite algebraic extension of k. (This is another version of Hilbert's Nullstellensatz. The following proof is due to Zariski. For other proofs, see (5.24), (7.9).)

Let $x_1, \ldots, x_n$ generate B as a k-algebra. The proof is by induction on n. If $n = 1$ the result is clearly true, so assume $n > 1$. Let $A = k[x_1]$ and let $K = k(x_1)$ be the field of fractions of A. By the inductive hypothesis, B is a finite algebraic extension of K, hence each of $x_2, \ldots, x_n$ satisfies a monic polynomial equation with coefficients in K, i.e. coefficients of the form a/b where a and b are in A. If f is the product of the denominators of all these coefficients, then each of $x_2, \ldots, x_n$ is integral over A_f. Hence B and therefore K is integral over A_f.

Suppose x_1 is transcendental over k. Then A is integrally closed, because it is a unique factorization domain. Hence A_f is integrally closed (5.12), and therefore $A_f = K$, which is clearly absurd. Hence x_1 is algebraic over k, hence K (and therefore B) is a finite extension of k.

19. Deduce the result of Exercise 17 from Exercise 18.

20. Let A be a subring of an integral domain B such that B is finitely generated over A. Show that there exists $s \neq 0$ in A and elements $y_1, \ldots, y_n$ in B, algebraically independent over A and such that B_s is integral over B'_s, where $B' = A[y_1, \ldots, y_n]$. [Let $S = A - \{0\}$ and let $K = S^{-1}A$, the field of fractions of A. Then $S^{-1}B$ is a finitely generated K-algebra and therefore by the normalization lemma (Exercise 16) there exist $x_1, \ldots, x_n$ in $S^{-1}B$, algebraically independent over K and such that $S^{-1}B$ is integral over $K[x_1, \ldots, x_n]$. Let $z_1, \ldots, z_m$ generate B as an A-algebra. Then each z_j (regarded as an element of $S^{-1}B$) is integral over $K[x_1, \ldots, x_n]$. By writing an equation of integral dependence for each z_j, show that there exists $s \in S$ such that $x_i = y_i/s$ $(1 \leqslant i \leqslant n)$ with $y_i \in B$, and such that each sz_j is integral over B'. Deduce that this s satisfies the conditions stated.]

21. Let A, B be as in Exercise 20. Show that there exists $s \neq 0$ in A such that, if Ω is an algebraically closed field and $f: A \to \Omega$ is a homomorphism for which $f(s) \neq 0$, then f can be extended to a homomorphism $B \to \Omega$. [With the notation of Exercise 20, f can be extended first of all to B', for example by mapping each y_i to 0; then to B'_s (because $f(s) \neq 0$), and finally to B_s (by Exercise 2, because B_s is integral over B'_s).]

22. Let A, B be as in Exercise 20. If the Jacobson radical of A is zero, then so is the Jacobson radical of B.
[Ley $v \neq 0$ be an element of B. We have to show that there is a maximal ideal of B which does not contain v. By applying Exercise 21 to the ring B_v and its subring A, we obtain an element $s \neq 0$ in A. Let $\mathfrak{m}$ be a maximal ideal of A such that $s \notin \mathfrak{m}$, and let $k = A/\mathfrak{m}$. Then the canonical mapping $A \to k$ extends to a homomorphism g of B_v into an algebraic closure Ω of k. Show that $g(v) \neq 0$ and that $\operatorname{Ker}(g) \cap B$ is a maximal ideal of B.]

23. Let A be a ring. Show that the following are equivalent:

i) Every prime ideal in A is an intersection of maximal ideals.

ii) In every homomorphic image of A the nilradical is equal to the Jacobson radical.

iii) Every prime ideal in A which is not maximal is equal to the intersection of the prime ideals which contain it strictly.

[The only hard part is iii) $\Rightarrow$ ii). Suppose ii) false, then there is a prime ideal which is not an intersection of maximal ideals. Passing to the quotient ring, we may assume that A is an integral domain whose Jacobson radical $\mathfrak{R}$ is not zero. Let f be a non-zero element of $\mathfrak{R}$. Then $A_f \neq 0$, hence A_f has a maximal ideal, whose contraction in A is a prime ideal $\mathfrak{p}$ such that $f \notin \mathfrak{p}$, and which is maximal with respect to this property. Then $\mathfrak{p}$ is not maximal and is not equal to the intersection of the prime ideals strictly containing $\mathfrak{p}$.]

A ring A with the three equivalent properties above is called a *Jacobson ring*.

24. Let A be a Jacobson ring (Exercise 23) and B an A-algebra. Show that if B is either (i) integral over A or (ii) finitely generated as an A-algebra, then B is Jacobson. [Use Exercise 22 for (ii).]

In particular, every finitely generated ring, and every finitely generated algebra over a field, is a Jacobson ring.

25. Let A be a ring. Show that the following are equivalent:

i) A is a Jacobson ring;

ii) Every finitely generated A-algebra B which is a field is finite over A.

[i) $\Rightarrow$ ii). Reduce to the case where A is a subring of B, and use Exercise 21. If $s \in A$ is as in Exercise 21, then there exists a maximal ideal $\mathfrak{m}$ of A not containing s, and the homomorphism $A \to A/\mathfrak{m} = k$ extends to a homomorphism g of B into the algebraic closure of k. Since B is a field, g is injective, and $g(B)$ is algebraic over k, hence finite algebraic over k.

ii) $\Rightarrow$ i). Use criterion iii) of Exercise 23. Let $\mathfrak{p}$ be a prime ideal of A which is not maximal, and let $B = A/\mathfrak{p}$. Let f be a non-zero element of B. Then B_f is a finitely generated A-algebra. If it is a field it is finite over B, hence integral over B and therefore B is a field by (5.7). Hence B_f is not a field and therefore has a non-zero prime ideal, whose contraction in B is a non-zero ideal $\mathfrak{p}'$ such that $f \notin \mathfrak{p}'$.]

26. Let X be a topological space. A subset of X is *locally closed* if it is the intersection of an open set and a closed set, or equivalently if it is open in its closure.

The following conditions on a subset X_0 of X are equivalent:

(1) Every non-empty locally closed subset of X meets X_0;

(2) For every closed set E in X we have $\overline{E \cap X_0} = E$;

(3) The mapping $U \mapsto U \cap X_0$ of the collection of open sets of X onto the collection of open sets of X_0 is *bijective*.

A subset X_0 satisfying these conditions is said to be *very dense* in X.

If A is a ring, show that the following are equivalent:

i) A is a Jacobson ring;

ii) The set of maximal ideals of A is very dense in Spec (A);

iii) Every locally closed subset of Spec (A) consisting of a single point is closed.
[ii) and iii) are geometrical formulations of conditions ii) and iii) of Exercise 23.]

Valuation rings and valuations

27. Let A, B be two local rings. B is said to *dominate* A if A is a subring of B and the maximal ideal $\mathfrak{m}$ of A is contained in the maximal ideal $\mathfrak{n}$ of B (or, equivalently, if $\mathfrak{m} = \mathfrak{n} \cap A$). Let K be a field and let Σ be the set of all local subrings of K. If Σ is ordered by the relation of domination, show that Σ has maximal elements and that $A \in \Sigma$ is maximal if and only if A is a valuation ring of K. [Use (5.21).]

28. Let A be an integral domain, K its field of fractions. Show that the following are equivalent:
(1) A is a valuation ring of K;
(2) If $\mathfrak{a}$, $\mathfrak{b}$ are any two ideals of A, then either $\mathfrak{a} \subseteq \mathfrak{b}$ or $\mathfrak{b} \subseteq \mathfrak{a}$.
Deduce that if A is a valuation ring and $\mathfrak{p}$ is a prime ideal of A, then $A_{\mathfrak{p}}$ and $A/\mathfrak{p}$ are valuation rings of their fields of fractions.

29. Let A be a valuation ring of a field K. Show that every subring of K which contains A is a local ring of A.

30. Let A be a valuation ring of a field K. The group U of units of A is a subgroup of the multiplicative group K^* of K.
Let $\Gamma = K^*/U$. If $\xi, \eta \in \Gamma$ are represented by $x, y \in K$, define $\xi \geqslant \eta$ to mean $xy^{-1} \in A$. Show that this defines a total ordering on Γ which is compatible with the group structure (i.e., $\xi \geqslant \eta \Rightarrow \xi\omega \geqslant \eta\omega$ for all $\omega \in \Gamma$). In other words, Γ is a totally ordered abelian group. It is called the *value group* of A.
Let $v: K^* \to \Gamma$ be the canonical homomorphism. Show that $v(x + y) \geqslant \min(v(x), v(y))$ for all $x, y \in K^*$.

31. Conversely, let Γ be a totally ordered abelian group (written *additively*), and let K be a field. A *valuation of K with values in* Γ is a mapping $v: K^* \to \Gamma$ such that
(1) $v(xy) = v(x) + v(y)$,
(2) $v(x + y) \geqslant \min(v(x), v(y))$,
for all $x, y \in K^*$. Show that the set of elements $x \in K^*$ such that $v(x) \geqslant 0$ is a valuation ring of K. This ring is called the *valuation ring* of v, and the subgroup $v(K^*)$ of Γ is the *value group* of v.
Thus the concepts of valuation ring and valuation are essentially equivalent.

32. Let Γ be a totally ordered abelian group. A subgroup Δ of Γ is *isolated* in Γ if, whenever $0 \leqslant \beta \leqslant \alpha$ and $\alpha \in \Delta$, we have $\beta \in \Delta$. Let A be a valuation ring of a field K, with value group Γ (Exercise 31). If $\mathfrak{p}$ is a prime ideal of A, show that $v(A - \mathfrak{p})$ is the set of elements $\geqslant 0$ in an isolated subgroup Δ of Γ, and that the mapping so defined of Spec (A) into the set of isolated subgroups of Γ is bijective.
If $\mathfrak{p}$ is a prime ideal of A, what are the value groups of the valuation rings $A/\mathfrak{p}$, $A_{\mathfrak{p}}$?

33. Let Γ be a totally ordered abelian group. We shall show how to construct a field K and a valuation v of K with Γ as value group. Let k be any field and let

$A = k[\Gamma]$ be the group algebra of Γ over k. By definition, A is freely generated as a k-vector space by elements x_α ($\alpha \in \Gamma$) such that $x_\alpha x_\beta = x_{\alpha+\beta}$. Show that A is an integral domain.

If $u = \lambda_1 x_{\alpha_1} + \cdots + \lambda_n x_{\alpha_n}$ is any non-zero element of A, where the λ_i are all $\neq 0$ and $\alpha_1 < \cdots < \alpha_n$, define $v_0(u)$ to be α_1. Show that the mapping $v_0: A - \{0\} \to \Gamma$ satisfies conditions (1) and (2) of Exercise 31.

Let K be the field of fractions of A. Show that v_0 can be uniquely extended to a valuation v of K, and that the value group of v is precisely Γ.

34. Let A be a valuation ring and K its field of fractions. Let $f: A \to B$ be a ring homomorphism such that f^*: Spec $(B) \to$ Spec (A) is a *closed* mapping. Then if $g: B \to K$ is any A-algebra homomorphism (i.e., if $g \circ f$ is the embedding of A in K) we have $g(B) = A$.
[Let $C = g(B)$; obviously $C \supseteq A$. Let $\mathfrak{n}$ be a maximal ideal of C. Since f^* is closed, $\mathfrak{m} = \mathfrak{n} \cap A$ is the maximal ideal of A, whence $A_\mathfrak{m} = A$. Also the local ring $C_\mathfrak{n}$ dominates $A_\mathfrak{m}$. Hence by Exercise 27 we have $C_\mathfrak{n} = A$ and therefore $C \subseteq A$.]

35. From Exercises 1 and 3 it follows that, if $f: A \to B$ is integral and C is any A-algebra, then the mapping $(f \otimes 1)^*$: Spec $(B \otimes_A C) \to$ Spec (C) is a closed map.

Conversely, suppose that $f: A \to B$ has this property and that B is an integral domain. Then f is integral. [Replacing A by its image in B, reduce to the case where $A \subseteq B$ and f is the injection. Let K be the field of fractions of B and let A' be a valuation ring of K containing A. By (5.22) it is enough to show that A' contains B. By hypothesis Spec $(B \otimes_A A') \to$ Spec (A') is a closed map. Apply the result of Exercise 34 to the homomorphism $B \otimes_A A' \to K$ defined by $b \otimes a' \mapsto ba'$. It follows that $ba' \in A'$ for all $b \in B$ and all $a' \in A'$; taking $a' = 1$, we have what we want.]

Show that the result just proved remains valid if B is a ring with only finitely many minimal prime ideals (e.g., if B is Noetherian). [Let $\mathfrak{p}_i$ be the minimal prime ideals. Then each composite homomorphism $A \to B \to B/\mathfrak{p}_i$ is integral, hence $A \to \Pi\,(B/\mathfrak{p}_i)$ is integral, hence $A \to B/\mathfrak{N}$ is integral (where $\mathfrak{N}$ is the nilradical of B), hence finally $A \to B$ is integral.]

6

Chain Conditions

So far we have considered quite arbitrary commutative rings (with identity). To go further, however, and obtain deeper theorems we need to impose some finiteness conditions. The most convenient way is in the form of "chain conditions". These apply both to rings and modules, and in this chapter we consider the case of modules. Most of the arguments are of a rather formal kind and because of this there is a symmetry between the ascending and descending chains—a symmetry which disappears in the case of rings as we shall see in subsequent chapters.

Let Σ be a set partially ordered by a relation $\leqslant$ (i.e., $\leqslant$ is reflexive and transitive and is such that $x \leqslant y$ and $y \leqslant x$ together imply $x = y$).

Proposition 6.1. *The following conditions on* Σ *are equivalent:*

i) *Every increasing sequence* $x_1 \leqslant x_2 \leqslant \cdots$ *in* Σ *is stationary (i.e., there exists n such that* $x_n = x_{n+1} = \cdots$*).*

ii) *Every non-empty subset of* Σ *has a maximal element.*

Proof. i) ⇒ ii). If ii) is false there is a non-empty subset T of Σ with no maximal element, and we can construct inductively a non-terminating strictly increasing sequence in T.

ii) ⇒ i). The set $(x_m)_{m \geq 1}$ has a maximal element, say x_n. ∎

If Σ is the set of submodules of a module M, ordered by the relation $\subseteq$, then i) is called the *ascending chain condition* (a.c.c. for short) and ii) the *maximal condition*. A module M satisfying either of these equivalent conditions is said to be *Noetherian* (after Emmy Noether). If Σ is ordered by $\supseteq$, then i) is the *descending chain condition* (d.c.c. for short) and ii) the *minimal condition*. A module M satisfying these is said to be *Artinian* (after Emil Artin).

Examples. 1) A finite abelian group (as **Z**-module) satisfies both a.c.c. and d.c.c.

2) The ring **Z** (as **Z**-module) satisfies a.c.c. but not d.c.c. For if $a \in \mathbf{Z}$ and $a \neq 0$ we have $(a) \supset (a^2) \supset \cdots \supset (a^n) \supset \cdots$ (strict inclusions).

3) Let G be the subgroup of **Q**/**Z** consisting of all elements whose order is a power of p, where p is a fixed prime. Then G has exactly one subgroup G_n of

order p^n for each $n \geqslant 0$, and $G_0 \subset G_1 \subset \cdots \subset G_n \subset \cdots$ (strict inclusions) so that G does not satisfy the a.c.c. On the other hand the only proper subgroups of G are the G_n, so that G does satisfy d.c.c.

4) The group H of all rational numbers of the form m/p^n ($m, n \in \mathbf{Z}, n \geqslant 0$) satisfies neither chain condition. For we have an exact sequence $0 \to \mathbf{Z} \to H \to G \to 0$, so that H doesn't satisfy d.c.c. because $\mathbf{Z}$ doesn't; and H doesn't satisfy a.c.c. because G doesn't.

5) The ring $k[x]$ (k a field, x an indeterminate) satisfies a.c.c. but not d.c.c. on *ideals*.

6) The polynomial ring $k[x_1, x_2, \ldots]$ in an infinite number of indeterminates x_n satisfies *neither* chain condition on ideals: for the sequence $(x_1) \subset (x_1, x_2) \subset \cdots$ is strictly increasing, and the sequence $(x_1) \supset (x_1^2) \supset (x_1^3) \supset \cdots$ is strictly decreasing.

7) We shall see later that a ring which satisfies d.c.c. on ideals must also satisfy a.c.c. on ideals. (This is *not* true in general for modules: see Examples 2, 3 above.)

Proposition 6.2. *M is a Noetherian A-module $\Leftrightarrow$ every submodule of M is finitely generated.*

Proof. $\Rightarrow$: Let N be a submodule of M, and let Σ be the set of all finitely generated submodules of N. Then Σ is not empty (since $0 \in \Sigma$) and therefore has a maximal element, say N_0. If $N_0 \neq N$, consider the submodule $N_0 + Ax$ where $x \in N$, $x \notin N_0$; this is finitely generated and strictly contains N_0, so we have a contradiction. Hence $N = N_0$ and therefore N is finitely generated.

$\Leftarrow$: Let $M_1 \subseteq M_2 \subseteq \cdots$ be an ascending chain of submodules of M. Then $N = \bigcup_{n=1}^{\infty} M_n$ is a submodule of M, hence is finitely generated, say by $x_1, \ldots, x_r$. Say $x_i \in M_{n_i}$ and let $n = \max_{i=1}^{r} n_i$; then each $x_i \in M_n$, hence $M_n = M$ and therefore the chain is stationary. ∎

Because of (6.2), Noetherian modules are more important than Artinian modules: the Noetherian condition is just the right finiteness condition to make a lot of theorems work. However, many of the elementary formal properties apply equally to Noetherian and Artinian modules.

Proposition 6.3. *Let $0 \to M' \xrightarrow{\alpha} M \xrightarrow{\beta} M'' \to 0$ be an exact sequence of A-modules. Then*

i) *M is Noetherian $\Leftrightarrow$ M' and M'' are Noetherian;*

ii) *M is Artinian $\Leftrightarrow$ M' and M'' are Artinian.*

Proof. We shall prove i); the proof of ii) is similar.

$\Rightarrow$: An ascending chain of submodules of M' (or M'') gives rise to a chain in M, hence is stationary.

$\Leftarrow$: Let $(L_n)_{n \geq 1}$ be an ascending chain of submodules of M; then $(\alpha^{-1}(L_n))$ is a chain in M', and $(\beta(L_n))$ is a chain in M''. For large enough n both these chains are stationary, and it follows that the chain (L_n) is stationary. ∎

Corollary 6.4. If M_i $(1 \leq i \leq n)$ are Noetherian (resp. Artinian) A-modules, so is $\bigoplus_{i=1}^n M_i$.

Proof. Apply induction and (6.3) to the exact sequence

$$0 \to M_n \to \bigoplus_{i=1}^{n} M_i \to \bigoplus_{i=1}^{n-1} M_i \to 0. \quad ∎$$

A ring A is said to be *Noetherian* (resp. *Artinian*) if it is so as an A-module, i.e., if it satisfies a.c.c. (resp. d.c.c.) on *ideals.*

Examples. 1) Any field is both Artinian and Noetherian; so is the ring $\mathbf{Z}/(n)$ $(n \neq 0)$. The ring $\mathbf{Z}$ is Noetherian, but not Artinian (Exercise 2 before (6.2)).

2) Any principal ideal domain is Noetherian (by (6.2): every ideal is finitely generated).

3) The ring $k[x_1, x_2, \ldots]$ is not Noetherian (Exercise 6 above). But it is an integral domain, hence has a field of fractions. Thus a *subring* of a Noetherian ring need not be Noetherian.

4) Let X be a compact infinite Hausdorff space, $C(X)$ the ring of real-valued continuous functions on X. Take a strictly decreasing sequence $F_1 \supset F_2 \supset \cdots$ of closed sets in X, and let $\mathfrak{a}_n = \{f \in C(X) : f(F_n) = 0\}$. Then the $\mathfrak{a}_n$ form a strictly increasing sequence of ideals in $C(X)$: so $C(X)$ is not a Noetherian ring.

Proposition 6.5. Let A be a Noetherian (resp. *Artinian*) *ring, M a finitely-generated A-module. Then M is Noetherian* (resp. *Artinian*).

Proof. M is a quotient of A^n for some n: apply (6.4) and (6.3). ∎

Proposition 6.6. Let A be Noetherian (resp. *Artinian*), $\mathfrak{a}$ *an ideal of A. Then $A/\mathfrak{a}$ is a Noetherian* (resp. *Artinian*) *ring.*

Proof. By (6.3) $A/\mathfrak{a}$ is Noetherian (resp. Artinian) as an A-module, hence also as an $A/\mathfrak{a}$-module. ∎

A *chain* of submodules of a module M is a sequence (M_i) $(0 \leq i \leq n)$ of submodules of M such that

$$M = M_0 \supset M_1 \supset \cdots \supset M_n = 0 \text{ (\textit{strict} inclusions).}$$

The *length* of the chain is n (the number of "links"). A *composition series* of M is a maximal chain, that is one in which no extra submodules can be inserted: this is equivalent to saying that each quotient M_{i-1}/M_i $(1 \leq i \leq n)$ is *simple* (that is, has no submodules except 0 and itself).

Proposition 6.7. *Suppose that M has a composition series of length n. Then every composition series of M has length n, and every chain in M can be extended to a composition series.*

Proof. Let $l(M)$ denote the least length of a composition series of a module M. ($l(M) = +\infty$ if M has no composition series.)

i) $N \subset M \Rightarrow l(N) < l(M)$. Let (M_i) be a composition series of M of minimum length, and consider the submodules $N_i = N \cap M_i$ of N. Since $N_{i-1}/N_i \subseteq M_{i-1}/M_i$ and the latter is a simple module, we have either $N_{i-1}/N_i = M_{i-1}/M_i$, or else $N_{i-1} = N_i$; hence, removing repeated terms, we have a composition series of N, so that $l(N) \leqslant l(M)$. If $l(N) = l(M) = n$, then $N_{i-1}/N_i = M_{i-1}/M_i$ for each $i = 1, 2, \ldots, n$; hence $M_{n-1} = N_{n-1}$, hence $M_{n-2} = N_{n-2}, \ldots$, and finally $M = N$.

ii) *Any chain in M has length $\leqslant l(M)$.* Let $M = M_0 \supset M_1 \supset \cdots$ be a chain of length k. Then by i) we have $l(M) > l(M_1) > \cdots > l(M_k) = 0$, hence $l(M) \geqslant k$.

iii) Consider any composition series of M. If it has length k, then $k \leqslant l(M)$ by ii), hence $k = l(M)$ by the definition of $l(M)$. Hence all composition series have the same length. Finally, consider any chain. If its length is $l(M)$ it must be a composition series, by ii); if its length is $< l(M)$ it is not a composition series, hence not maximal, and therefore new terms can be inserted until the length is $l(M)$. ∎

Proposition 6.8. *M has a composition series $\Leftrightarrow$ M satisfies both chain conditions.*

Proof. $\Rightarrow$: All chains in M are of bounded length, hence both a.c.c. and d.c.c. hold.

$\Leftarrow$: Construct a composition series of M as follows. Since $M = M_0$ satisfies the maximum condition by (6.1), it has a maximal submodule $M_1 \subset M_0$. Similarly M_1 has a maximal submodule $M_2 \subset M_1$, and so on. Thus we have a strictly descending chain $M_0 \supset M_1 \supset \cdots$ which by d.c.c. must be finite, and hence is a composition series of M. ∎

A module satisfying both a.c.c. and d.c.c. is therefore called a *module of finite length*. By (6.7) all composition series of M have the same length $l(M)$, called the *length of M*. The Jordan–Hölder theorem applies to modules of finite length: if $(M_i)_{0 \leqslant i \leqslant n}$ and $(M'_i)_{0 \leqslant i \leqslant n}$ are any two composition series of M, there is a one-to-one correspondence between the set of quotients $(M_{i-1}/M_i)_{1 \leqslant i \leqslant n}$ and the set of quotients $(M'_{i-1}/M'_i)_{1 \leqslant i \leqslant n}$, such that corresponding quotients are isomorphic. The proof is the same as for finite groups.

Proposition 6.9. *The length $l(M)$ is an additive function on the class of all A-modules of finite length.*

Proof. We have to show that if $0 \rightarrow M' \xrightarrow{\alpha} M \xrightarrow{\beta} M'' \rightarrow 0$ is an exact sequence, then $l(M) = l(M') + l(M'')$. Take the image under α of any composition series of M' and the inverse image under β of any composition series of M''; these fit together to give a composition series of M, hence the result. ∎

Consider the particular case of modules over a field k, i.e., k-vector spaces:

Proposition 6.10. *For k-vector spaces V the following conditions are equivalent:*

i) *finite dimension;*

ii) *finite length;*

iii) *a.c.c.;*

iv) *d.c.c.*

Moreover, if these conditions are satisfied, length = dimension.

Proof. i) ⇒ ii) is elementary; ii) ⇒ iii), ii) ⇒ iv) from (6.8). Remains to prove iii) ⇒ i) and iv) ⇒ i). Suppose i) is false, then there exists an infinite sequence $(x_n)_{n\geq 1}$ of linearly independent elements of V. Let U_n (resp. V_n) be the vector space spanned by $x_1, \ldots, x_n$ (resp., $x_{n+1}, x_{n+2}, \ldots$). Then the chain $(U_n)_{n\geq 1}$ (resp. $(V_n)_{n\geq 1}$) is infinite and strictly ascending (resp. strictly descending).

Corollary 6.11. *Let A be a ring in which the zero ideal is a product* $\mathfrak{m}_1 \cdots \mathfrak{m}_n$ *of (not necessarily distinct) maximal ideals. Then A is Noetherian if and only if A is Artinian.*

Proof. Consider the chain of ideals $A \supset \mathfrak{m}_1 \supseteq \mathfrak{m}_1\mathfrak{m}_2 \supseteq \cdots \supseteq \mathfrak{m}_1 \cdots \mathfrak{m}_n = 0$. Each factor $\mathfrak{m}_1 \cdots \mathfrak{m}_{i-1}/\mathfrak{m}_1 \cdots \mathfrak{m}_i$ is a vector space over the field $A/\mathfrak{m}_i$. Hence a.c.c. ⇔ d.c.c. for each factor. But a.c.c. (resp. d.c.c.) for each factor ⇔ a.c.c. (resp. d.c.c.) for A, by repeated application of (6.3). Hence a.c.c. ⇔ d.c.c. for A. ∎

EXERCISES

1. i) Let M be a Noetherian A-module and $u: M \to M$ a module homomorphism. If u is surjective, then u is an isomorphism.

   ii) If M is Artinian and u is injective, then again u is an isomorphism.

   [For (i), consider the submodules $\operatorname{Ker}(u^n)$; for (ii), the quotient modules $\operatorname{Coker}(u^n)$.]

2. Let M be an A-module. If every non-empty set of finitely generated submodules of M has a maximal element, then M is Noetherian.

3. Let M be an A-module and let N_1, N_2 be submodules of M. If M/N_1 and M/N_2 are Noetherian, so is $M/(N_1 \cap N_2)$. Similarly with Artinian in place of Noetherian.

4. Let M be a Noetherian A-module and let $\mathfrak{a}$ be the annihilator of M in A. Prove that $A/\mathfrak{a}$ is a Noetherian ring.

   If we replace "Noetherian" by "Artinian" in this result, is it still true?

5. A topological space X is said to be *Noetherian* if the open subsets of X satisfy the ascending chain condition (or, equivalently, the maximal condition). Since closed subsets are complements of open subsets, it comes to the same thing to say that the closed subsets of X satisfy the descending chain condition (or, equivalently, the minimal condition). Show that, if X is Noetherian, then every subspace of X is Noetherian, and that X is quasi-compact.

6. Prove that the following are equivalent:
   i) X is Noetherian.
   ii) Every open subspace of X is quasi-compact.
   iii) Every subspace of X is quasi-compact.

7. A Noetherian space is a finite union of irreducible closed subspaces. [Consider the set Σ of closed subsets of X which are not finite unions of irreducible closed subspaces.] Hence the set of irreducible components of a Noetherian space is finite.

8. If A is a Noetherian ring then Spec (A) is a Noetherian topological space. Is the converse true?

9. Deduce from Exercise 8 that the set of minimal prime ideals in a Noetherian ring is finite.

10. If M is a Noetherian module (over an arbitrary ring A) then Supp (M) is a closed Noetherian subspace of Spec (A).

11. Let $f: A \to B$ be a ring homomorphism and suppose that Spec (B) is a Noetherian space (Exercise 5). Prove that f^*: Spec (B) $\to$ Spec (A) is a closed mapping if and only if f has the going-up property (Chapter 5, Exercise 10).

12. Let A be a ring such that Spec (A) is a Noetherian space. Show that the set of prime ideals of A satisfies the ascending chain condition. Is the converse true?

7

Noetherian Rings

We recall that a ring A is said to be *Noetherian* if it satisfies the following three equivalent conditions:

1) Every non-empty set of ideals in A has a maximal element.
2) Every ascending chain of ideals in A is stationary.
3) Every ideal in A is finitely generated.

(The equivalence of these conditions was proved in (6.1) and (6.2).)

Noetherian rings are by far the most important class of rings in commutative algebra: we have seen some examples already in Chapter 6. In this chapter we shall first show that Noetherian rings reproduce themselves under various familiar operations—in particular we prove the famous basis theorem of Hilbert. We then proceed to make a number of important deductions from the Noetherian condition, including the existence of primary decompositions.

***Proposition** 7.1. If A is Noetherian and ϕ is a homomorphism of A onto a ring B, then B is Noetherian.*

Proof. This follows from (6.6), since $B \cong A/\mathfrak{a}$, where $\mathfrak{a} = \operatorname{Ker}(\phi)$. ■

***Proposition** 7.2. Let A be a subring of B; suppose that A is Noetherian and that B is finitely generated as an A-module. Then B is Noetherian (as a ring).*

Proof. By (6.5) B is Noetherian as an A-module, hence also as a B-module. ■

Example. $B = \mathbf{Z}[i]$, the ring of Gaussian integers. By (7.2) B is Noetherian. More generally, the ring of integers in any algebraic number field is Noetherian.

***Proposition** 7.3. If A is Noetherian and S is any multiplicatively closed subset of A, then $S^{-1}A$ is Noetherian.*

Proof. By (3.11—i) and (1.17—iii) the ideals of $S^{-1}A$ are in one-to-one order-preserving correspondence with the contracted ideals of A, hence satisfy the maximal condition. (Alternative proof: if $\mathfrak{a}$ is any ideal of A, then $\mathfrak{a}$ has a finite set of generators, say $x_1, \ldots, x_n$, and it is clear that $S^{-1}\mathfrak{a}$ is generated by $x_1/1, \ldots, x_n/1$.) ■

***Corollary** 7.4. If A is Noetherian and $\mathfrak{p}$ is a prime ideal of A, then $A_\mathfrak{p}$ is Noetherian.* ■

Theorem 7.5. (Hilbert's Basis Theorem). *If A is Noetherian, then the polynomial ring $A[x]$ is Noetherian.*

Proof. Let $\mathfrak{a}$ be an ideal in $A[x]$. The leading coefficients of the polynomials in $\mathfrak{a}$ form an ideal $\mathfrak{l}$ in A. Since A is Noetherian, $\mathfrak{l}$ is finitely generated, say by $a_1, \ldots, a_n$. For each $i = 1, \ldots, n$ there is a polynomial $f_i \in A[x]$ of the form $f_i = a_i x^{r_i} +$ (lower terms). Let $r = \max_{i=1}^n r_i$. The f_i generate an ideal $\mathfrak{a}' \subseteq \mathfrak{a}$ in $A[x]$.

Let $f = ax^m +$ (lower terms) be any element of $\mathfrak{a}$; we have $a \in \mathfrak{l}$. If $m \geqslant r$, write $a = \sum_{i=1}^n u_i a_i$, where $u_i \in A$; then $f - \sum u_i f_i x^{m-r_i}$ is in $\mathfrak{a}$ and has degree $< m$. Proceeding in this way, we can go on subtracting elements of $\mathfrak{a}'$ from f until we get a polynomial g, say, of degree $< r$; that is, we have $f = g + h$, where $h \in \mathfrak{a}'$.

Let M be the A-module generated by $1, x, \ldots, x^{r-1}$; then what we have proved is that $\mathfrak{a} = (\mathfrak{a} \cap M) + \mathfrak{a}'$. Now M is a finitely generated A-module, hence is Noetherian by (6.5), hence $\mathfrak{a} \cap M$ is finitely generated (as an A-module) by (6.2). If $g_1, \ldots, g_m$ generate $\mathfrak{a} \cap M$ it is clear that the f_i and the g_j generate $\mathfrak{a}$. Hence $\mathfrak{a}$ is finitely generated and so $A[x]$ is Noetherian. ∎

Remark. It is also true that A Noetherian $\Rightarrow A[[x]]$ Noetherian ($A[[x]]$ being the ring of formal power series in x with coefficients in A). The proof runs almost parallel to that of (7.5) except that one starts with the terms of *lowest* degree in the power series belonging to $\mathfrak{a}$. See also (10.27).

Corollary 7.6. If A is Noetherian so is $A[x_1, \ldots, x_n]$.

Proof. By induction on n from (7.5). ∎

Corollary 7.7. Let B be a finitely-generated A-algebra. If A is Noetherian, then so is B.

In particular, every finitely-generated ring, and every finitely generated algebra over a field, is Noetherian.

Proof. B is a homomorphic image of a polynomial ring $A[x_1, \ldots, x_n]$, which is Noetherian by (7.6). ∎

Proposition 7.8. Let $A \subseteq B \subseteq C$ be rings. Suppose that A is Noetherian, that C is finitely generated as an A-algebra and that C is either (i) *finitely generated as a B-module or* (ii) *integral over B. Then B is finitely generated as an A-algebra.*

Proof. It follows from (5.1) and (5.2) that the conditions (i) and (ii) are equivalent in this situation. So we may concentrate on (i).

Let $x_1, \ldots, x_m$ generate C as an A-algebra, and let $y_1, \ldots, y_n$ generate C as a B-module. Then there exist expressions of the form

$$x_i = \sum_j b_{ij} y_j \qquad (b_{ij} \in B) \tag{1}$$

$$y_i y_j = \sum_k b_{ijk} y_k \qquad (b_{ijk} \in B). \tag{2}$$

Let B_0 be the algebra generated over A by the b_{ij} and the b_{ijk}. Since A is Noetherian, so is B_0 by (7.7), and $A \subseteq B_0 \subseteq B$.

Any element of C is a polynomial in the x_i with coefficients in A. Substituting (1) and making repeated use of (2) shows that each element of C is a linear combination of the y_j with coefficients in B_0, and hence C is finitely generated as a B_0-module. Since B_0 is Noetherian, and B is a submodule of C, it follows (by (6.5) and (6.2)) that B is finitely generated as a B_0-module. Since B_0 is finitely generated as an A-algebra, it follows that B is finitely-generated as an A-algebra. ∎

***Proposition** 7.9. Let k be a field, E a finitely generated k-algebra. If E is a field then it is a finite algebraic extension of k.*

Proof. Let $E = k[x_1, \ldots, x_n]$. If E is not algebraic over k then we can renumber the x_i so that $x_1, \ldots, x_r$ are algebraically independent over k, where $r \geqslant 1$, and each of $x_{r+1}, \ldots, x_n$ is algebraic over the field $F = k(x_1, \ldots, x_r)$. Hence E is a finite algebraic extension of F and therefore finitely generated as an F-module. Applying (7.8) to $k \subseteq F \subseteq E$, it follows that F is a finitely generated k-algebra, say $F = k[y_1, \ldots, y_s]$. Each y_j is of the form f_j/g_j, where f_j and g_j are polynomials in $x_1, \ldots, x_r$.

Now there are infinitely many irreducible polynomials in the ring $k[x_1, \ldots, x_r]$ (adapt Euclid's proof of the existence of infinitely many prime numbers). Hence there is an irreducible polynomial h which is prime to each of the g_j (for example, $h = g_1 g_2 \cdots g_s + 1$ would do) and the element h^{-1} of F could not be a polynomial in the y_j. This is a contradiction. Hence E is algebraic over k, and therefore finite algebraic. ∎

***Corollary** 7.10. Let k be a field, A a finitely generated k-algebra. Let $\mathfrak{m}$ be a maximal ideal of A. Then the field $A/\mathfrak{m}$ is a finite algebraic extension of k. In particular, if k is algebraically closed then $A/\mathfrak{m} \cong k$.*

Proof. Take $E = A/\mathfrak{m}$ in (7.9). ∎

(7.10) is the so-called "weak" version of Hilbert's Nullstellensatz (= theorem of the zeros). The proof given here is due to Artin and Tate. For its geometrical meaning, and the "strong" form of the theorem, see the Exercises at the end of this chapter.

PRIMARY DECOMPOSITION IN NOETHERIAN RINGS

The next two lemmas show that every ideal $\neq (1)$ in a Noetherian ring has a primary decomposition.

An ideal $\mathfrak{a}$ is said to be *irreducible* if

$$\mathfrak{a} = \mathfrak{b} \cap \mathfrak{c} \Rightarrow (\mathfrak{a} = \mathfrak{b} \text{ or } \mathfrak{a} = \mathfrak{c}).$$

***Lemma* 7.11.** *In a Noetherian ring A every ideal is a finite intersection of irreducible ideals.*

Proof. Suppose not; then the set of ideals in A for which the lemma is false is not empty, hence has a maximal element $\mathfrak{a}$. Since $\mathfrak{a}$ is reducible, we have $\mathfrak{a} = \mathfrak{b} \cap \mathfrak{c}$ where $\mathfrak{b} \supset \mathfrak{a}$ and $\mathfrak{c} \supset \mathfrak{a}$. Hence each of $\mathfrak{b}$, $\mathfrak{c}$ is a finite intersection of irreducible ideals and therefore so is $\mathfrak{a}$: contradiction. ∎

***Lemma* 7.12.** *In a Noetherian ring every irreducible ideal is primary.*

Proof. By passing to the quotient ring, it is enough to show that if the zero ideal is irreducible then it is primary. Let $xy = 0$ with $y \neq 0$, and consider the chain of ideals $\text{Ann}(x) \subseteq \text{Ann}(x^2) \subseteq \cdots$. By the a.c.c., this chain is stationary, i.e., we have $\text{Ann}(x^n) = \text{Ann}(x^{n+1}) = \cdots$ for some n. It follows that $(x^n) \cap (y) = 0$; for if $a \in (y)$ then $ax = 0$, and if $a \in (x^n)$ then $a = bx^n$, hence $bx^{n+1} = 0$, hence $b \in \text{Ann}(x^{n+1}) = \text{Ann}(x^n)$, hence $bx^n = 0$; that is, $a = 0$. Since (0) is irreducible and $(y) \neq 0$ we must therefore have $x^n = 0$, and this shows that (0) is primary. ∎

From these two lemmas we have at once

***Theorem* 7.13.** *In a Noetherian ring A every ideal has a primary decomposition.* ∎

Hence all the results of Chapter 4 apply to Noetherian rings.

***Proposition* 7.14.** *In a Noetherian ring A, every ideal $\mathfrak{a}$ contains a power of its radical.*

Proof. Let $x_1, \ldots, x_k$ generate $r(\mathfrak{a})$: say $x_i^{n_i} \in \mathfrak{a}$ $(1 \leqslant i \leqslant k)$. Let $m = \sum_{i=1}^{k} (n_i - 1) + 1$. Then $r(\mathfrak{a})^m$ is generated by the products $x_1^{r_1} \cdots x_k^{r_k}$ with $\sum r_i = m$; from the definition of m we must have $r_i \geqslant n_i$ for at least one index i, hence each such monomial lies in $\mathfrak{a}$, and therefore $r(\mathfrak{a})^m \subseteq \mathfrak{a}$. ∎

***Corollary* 7.15.** *In a Noetherian ring the nilradical is nilpotent.*

Proof. Take $\mathfrak{a} = (0)$ in (7.14). ∎

***Corollary* 7.16.** *Let A be a Noetherian ring, $\mathfrak{m}$ a maximal ideal of A, $\mathfrak{q}$ any ideal of A. Then the following are equivalent:*

i) $\mathfrak{q}$ *is* $\mathfrak{m}$*-primary;*

ii) $r(\mathfrak{q}) = \mathfrak{m}$*;*

iii) $\mathfrak{m}^n \subseteq \mathfrak{q} \subseteq \mathfrak{m}$ *for some* $n > 0$.

Proof. i) ⇒ ii) is clear; ii) ⇒ i) from (4.2); ii) ⇒ iii) from (7.14); iii) ⇒ ii) by taking radicals: $\mathfrak{m} = r(\mathfrak{m}^n) \subseteq r(\mathfrak{q}) \subseteq r(\mathfrak{m}) = \mathfrak{m}$. ∎

***Proposition* 7.17.** *Let $\mathfrak{a} \neq (1)$ be an ideal in a Noetherian ring. Then the prime ideals which belong to $\mathfrak{a}$ are precisely the prime ideals which occur in the set of ideals $(\mathfrak{a}:x)$ $(x \in A)$.*

Proof. By passing to $A/\mathfrak{a}$ we may assume that $\mathfrak{a} = 0$. Let $\bigcap_{i=1}^{n} \mathfrak{q}_i = 0$ be a minimal primary decomposition of the zero ideal, and let $\mathfrak{p}_i$ be the radical of $\mathfrak{q}_i$.

Let $\mathfrak{a}_i = \bigcap_{j \neq i} \mathfrak{q}_j \neq 0$. Then from the proof of (4.5) we have $r(\text{Ann}(x)) = \mathfrak{p}_i$ for any $x \neq 0$ in $\mathfrak{a}_i$, so that $\text{Ann}(x) \subseteq \mathfrak{p}_i$.

Since $\mathfrak{q}_i$ is $\mathfrak{p}_i$-primary, by (7.14) there exists an integer m such that $\mathfrak{p}_i^m \subseteq \mathfrak{q}_i$, and therefore $\mathfrak{a}_i\mathfrak{p}_i^m \subseteq \mathfrak{a}_i \cap \mathfrak{p}_i^m \subseteq \mathfrak{a}_i \cap \mathfrak{q}_i = 0$. Let $m \geqslant 1$ be the smallest integer such that $\mathfrak{a}_i\mathfrak{p}_i^m = 0$, and let x be a non-zero element in $\mathfrak{a}_i\mathfrak{p}_i^{m-1}$. Then $\mathfrak{p}_i x = 0$, therefore for such an x we have $\text{Ann}(x) \supseteq \mathfrak{p}_i$, and hence $\text{Ann}(x) = \mathfrak{p}_i$.

Conversely, if $\text{Ann}(x)$ is a prime ideal $\mathfrak{p}$, then $r(\text{Ann}(x)) = \mathfrak{p}$ and so by (4.5) $\mathfrak{p}$ is a prime ideal belonging to 0. ∎

EXERCISES

1. Let A be a non-Noetherian ring and let Σ be the set of ideals in A which are not finitely generated. Show that Σ has maximal elements and that the maximal elements of Σ are prime ideals.
 [Let $\mathfrak{a}$ be a maximal element of Σ, and suppose that there exist $x, y \in A$ such that $x \notin \mathfrak{a}$ and $y \notin \mathfrak{a}$ and $xy \in \mathfrak{a}$. Show that there exists a finitely generated ideal $\mathfrak{a}_0 \subseteq \mathfrak{a}$ such that $\mathfrak{a}_0 + (x) = \mathfrak{a} + (x)$, and that $\mathfrak{a} = \mathfrak{a}_0 + x\cdot(\mathfrak{a}:x)$. Since $(\mathfrak{a}:x)$ strictly contains $\mathfrak{a}$, it is finitely generated and therefore so is $\mathfrak{a}$.]
 Hence a ring in which every prime ideal is finitely generated is Noetherian (I. S. Cohen).

2. Let A be a Noetherian ring and let $f = \sum_{n=0}^{\infty} a_n x^n \in A[[x]]$. Prove that f is nilpotent if and only if each a_n is nilpotent.

3. Let $\mathfrak{a}$ be an irreducible ideal in a ring A. Then the following are equivalent:
 i) $\mathfrak{a}$ is primary;
 ii) for every multiplicatively closed subset S of A we have $(S^{-1}\mathfrak{a})^c = (\mathfrak{a}:x)$ for some $x \in S$;
 iii) the sequence $(\mathfrak{a}:x^n)$ is stationary, for every $x \in A$.

4. Which of the following rings are Noetherian?
 i) The ring of rational functions of z having no pole on the circle $|z| = 1$.
 ii) The ring of power series in z with a positive radius of convergence.
 iii) The ring of power series in z with an infinite radius of convergence.
 iv) The ring of polynomials in z whose first k derivatives vanish at the origin (k being a fixed integer).
 v) The ring of polynomials in z, w all of whose partial derivatives with respect to w vanish for $z = 0$.
 In all cases the coefficients are complex numbers.

5. Let A be a Noetherian ring, B a finitely generated A-algebra, G a finite group of A-automorphisms of B, and B^G the set of all elements of B which are left fixed by every element of G. Show that B^G is a finitely generated A-algebra.

6. If a finitely generated ring K is a field, it is a finite field.
 [If K has characteristic 0, we have $\mathbf{Z} \subset \mathbf{Q} \subseteq K$. Since K is finitely generated over $\mathbf{Z}$ it is finitely generated over $\mathbf{Q}$, hence by (7.9) is a finitely generated $\mathbf{Q}$-

module. Now apply (7.8) to obtain a contradiction. Hence K is of characteristic $p > 0$, hence is finitely generated as a $\mathbf{Z}/(p)$-algebra. Use (7.9) to complete the proof.]

7. Let X be an affine algebraic variety given by a family of equations $f_\alpha(t_1, \ldots, t_n) = 0$ ($\alpha \in I$) (Chapter 1, Exercise 27). Show that there exists a finite subset I_0 of I such that X is given by the equations $f_\alpha(t_1, \ldots, t_n) = 0$ for $\alpha \in I_0$.

8. If $A[x]$ is Noetherian, is A necessarily Noetherian?

9. Let A be a ring such that
   (1) for each maximal ideal $\mathfrak{m}$ of A, the local ring $A_\mathfrak{m}$ is Noetherian;
   (2) for each $x \neq 0$ in A, the set of maximal ideals of A which contain x is finite.

   Show that A is Noetherian.
   [Let $\mathfrak{a} \neq 0$ be an ideal in A. Let $\mathfrak{m}_1, \ldots, \mathfrak{m}_r$ be the maximal ideals which contain $\mathfrak{a}$. Choose $x_0 \neq 0$ in $\mathfrak{a}$ and let $\mathfrak{m}_1, \ldots, \mathfrak{m}_{r+s}$ be the maximal ideals which contain x_0. Since $\mathfrak{m}_{r+1}, \ldots, \mathfrak{m}_{r+s}$ do not contain $\mathfrak{a}$ there exist $x_j \in \mathfrak{a}$ such that $x_j \notin \mathfrak{m}_{r+j}$ $(1 \leqslant j \leqslant s)$. Since each $A_{\mathfrak{m}_i}$ $(1 \leqslant i \leqslant r)$ is Noetherian, the extension of $\mathfrak{a}$ in $A_{\mathfrak{m}_i}$ is finitely generated. Hence there exist $x_{s+1}, \ldots, x_t$ in $\mathfrak{a}$ whose images in $A_{\mathfrak{m}_i}$ generate $A_{\mathfrak{m}_i}\mathfrak{a}$ for $i = 1, \ldots, r$. Let $\mathfrak{a}_0 = (x_0, \ldots, x_t)$. Show that $\mathfrak{a}_0$ and $\mathfrak{a}$ have the same extension in $A_\mathfrak{m}$ for every maximal ideal $\mathfrak{m}$, and deduce by (3.9) that $\mathfrak{a}_0 = \mathfrak{a}$.]

10. Let M be a Noetherian A-module. Show that $M[x]$ (Chapter 2, Exercise 6) is a Noetherian $A[x]$-module.

11. Let A be a ring such that each local ring $A_\mathfrak{p}$ is Noetherian. Is A necessarily Noetherian?

12. Let A be a ring and B a faithfully flat A-algebra (Chapter 3, Exercise 16). If B is Noetherian, show that A is Noetherian. [Use the ascending chain condition.]

13. Let $f: A \to B$ be a ring homomorphism of finite type and let $f^*: \operatorname{Spec}(B) \to \operatorname{Spec}(A)$ be the mapping associated with f. Show that the fibers of f^* are Noetherian subspaces of B.

Nullstellensatz, strong form

14. Let k be an algebraically closed field, let A denote the polynomial ring $k[t_1, \ldots, t_n]$ and let $\mathfrak{a}$ be an ideal in A. Let V be the variety in k^n defined by the ideal $\mathfrak{a}$, so that V is the set of all $x = (x_1, \ldots, x_n) \in k^n$ such that $f(x) = 0$ for all $f \in \mathfrak{a}$. Let $I(V)$ be the ideal of V, i.e. the ideal of all polynomials $g \in A$ such that $g(x) = 0$ for all $x \in V$. Then $I(V) = r(\mathfrak{a})$.
    [It is clear that $r(\mathfrak{a}) \subseteq I(V)$. Conversely, let $f \notin r(\mathfrak{a})$, then there is a prime ideal $\mathfrak{p}$ containing $\mathfrak{a}$ such that $f \notin \mathfrak{p}$. Let $\bar{f}$ be the image of f in $B = A/\mathfrak{p}$, let $C = B_f = B[1/\bar{f}]$, and let $\mathfrak{m}$ be a maximal ideal of C. Since C is a finitely generated k-algebra we have $C/\mathfrak{m} \cong k$, by (7.9). The images x_i in $C/\mathfrak{m}$ of the generators t_i of A thus define a point $x = (x_1, \ldots, x_n) \in k^n$, and the construction shows that $x \in V$ and $f(x) \neq 0$.]

15. Let A be a Noetherian local ring, $\mathfrak{m}$ its maximal ideal and k its residue field, and let M be a finitely generated A-module. Then the following are equivalent:

i) M is free;
ii) M is flat;
iii) the mapping of $\mathfrak{m} \otimes M$ into $A \otimes M$ is injective;
iv) $\mathrm{Tor}_1^A(k, M) = 0$.

[To show that iv) $\Rightarrow$ i), let $x_1, \ldots, x_n$ be elements of M whose images in $M/\mathfrak{m}M$ form a k-basis of this vector space. By (2.8), the x_i generate M. Let F be a free A-module with basis $e_1, \ldots, e_n$ and define $\phi: F \to M$ by $\phi(e_i) = x_i$. Let $E = \mathrm{Ker}(\phi)$. Then the exact sequence $0 \to E \to F \to M \to 0$ gives us an exact sequence

$$0 \longrightarrow k \otimes_A E \longrightarrow k \otimes_A F \xrightarrow{1 \otimes \phi} k \otimes_A M \longrightarrow 0.$$

Since $k \otimes F$ and $k \otimes M$ are vector spaces of the same dimension over k, it follows that $1 \otimes \phi$ is an isomorphism, hence $k \otimes E = 0$, hence $E = 0$ by Nakayama's Lemma (E is finitely generated because it is a submodule of F, and A is Noetherian).]

16. Let A be a Noetherian ring, M a finitely generated A-module. Then the following are equivalent:

i) M is a flat A-module;
ii) $M_\mathfrak{p}$ is a free $A_\mathfrak{p}$-module, for all prime ideals $\mathfrak{p}$;
iii) $M_\mathfrak{m}$ is a free $A_\mathfrak{m}$-module, for all maximal ideals $\mathfrak{m}$.

In other words, flat = locally free. [Use Exercise 15.]

17. Let A be a ring and M a Noetherian A-module. Show (by imitating the proofs of (7.11) and (7.12)) that every submodule N of M has a primary decomposition (Chapter 4, Exercises 20–23).

18. Let A be a Noetherian ring, $\mathfrak{p}$ a prime ideal of A, and M a finitely generated A-module. Show that the following are equivalent:

i) $\mathfrak{p}$ belongs to 0 in M;
ii) there exists $x \in M$ such that $\mathrm{Ann}(x) = \mathfrak{p}$;
iii) there exists a submodule of M isomorphic to $A/\mathfrak{p}$.

Deduce that there exists a chain of submodules

$$0 = M_0 \subset M_1 \subset \cdots \subset M_r = M$$

such that each quotient M_i/M_{i-1} is of the form $A/\mathfrak{p}_i$, where $\mathfrak{p}_i$ is a prime ideal of A.

19. Let $\mathfrak{a}$ be an ideal in a Noetherian ring A. Let

$$\mathfrak{a} = \bigcap_{i=1}^{r} \mathfrak{b}_i = \bigcap_{j=1}^{s} \mathfrak{c}_j$$

be two minimal decompositions of $\mathfrak{a}$ as intersections of *irreducible* ideals. Prove that $r = s$ and that (possible after re-indexing the $\mathfrak{c}_j$) $r(\mathfrak{b}_i) = r(\mathfrak{c}_i)$ for all i. [Show that for each $i = 1, \ldots, r$ there exists j such that

$$\mathfrak{a} = \mathfrak{b}_1 \cap \cdots \cap \mathfrak{b}_{i-1} \cap \mathfrak{c}_j \cap \mathfrak{b}_{i+1} \cap \cdots \cap \mathfrak{b}_r.]$$

State and prove an analogous result for modules.

20. Let X be a topological space and let $\mathscr{F}$ be the smallest collection of subsets of X which contains all open subsets of X and is closed with respect to the formation of finite intersections and complements.

i) Show that a subset E of X belongs to $\mathscr{F}$ if and only if E is a finite union of sets of the form $U \cap C$, where U is open and C is closed.

ii) Suppose that X is irreducible and let $E \in \mathscr{F}$. Show that E is dense in X (i.e., that $\bar{E} = X$) if and only if E contains a non-empty open set in X.

21. Let X be a Noetherian topological space (Chapter 6, Exercise 5) and let $E \subseteq X$. Show that $E \in \mathscr{F}$ if and only if, for each irreducible closed set $X_0 \subseteq X$, either $\overline{E \cap X_0} \neq X_0$ or else $E \cap X_0$ contains a non-empty open subset of X_0. [Suppose $E \notin \mathscr{F}$. Then the collection of closed sets $X' \subseteq X$ such that $E \cap X' \notin \mathscr{F}$ is not empty and therefore has a minimal element X_0. Show that X_0 is irreducible and then that each of the alternatives above leads to the conclusion that $E \cap X_0 \in \mathscr{F}$.] The sets belonging to $\mathscr{F}$ are called the *constructible* subsets of X.

22. Let X be a Noetherian topological space and let E be a subset of X. Show that E is open in X if and only if, for each irreducible closed subset X_0 in X, either $E \cap X_0 = \emptyset$ or else $E \cap X_0$ contains a non-empty open subset of X_0. [The proof is similar to that of Exercise 21.]

23. Let A be a Noetherian ring, $f: A \to B$ a ring homomorphism of finite type (so that B is Noetherian). Let $X = \operatorname{Spec}(A)$, $Y = \operatorname{Spec}(B)$ and let $f^*: Y \to X$ be the mapping associated with f. Then the image under f^* of a constructible subset E of Y is a constructible subset of X.
[By Exercise 20 it is enough to take $E = U \cap C$ where U is open and C is closed in Y; then, replacing B by a homomorphic image, we reduce to the case where E is open in Y. Since Y is Noetherian, E is quasi-compact and therefore a finite union of open sets of the form $\operatorname{Spec}(B_g)$. Hence reduce to the case $E = Y$. To show that $f^*(Y)$ is constructible, use the criterion of Exercise 21. Let X_0 be an irreducible closed subset of X such that $f^*(Y) \cap X_0$ is dense in X_0. We have $f^*(Y) \cap X_0 = f^*(f^{*-1}(X_0))$, and $f^{*-1}(X_0) = \operatorname{Spec}((A/\mathfrak{p}) \otimes_A B)$, where $X_0 = \operatorname{Spec}(A/\mathfrak{p})$. Hence reduce to the case where A is an integral domain and f is injective. If $Y_1, \ldots, Y_n$ are the irreducible components of Y, it is enough to show that some $f^*(Y_i)$ contains a non-empty open set in X. So finally we are brought down to the situation in which A, B are integral domains and f is injective (and still of finite type); now use Chapter 5, Exercise 21 to complete the proof.]

24. With the notation and hypotheses of Exercise 23, f^* is an open mapping $\Leftrightarrow$ f has the going-down property (Chapter 5, Exercise 10). [Suppose f has the going-down property. As in Exercise 23, reduce to proving that $E = f^*(Y)$ is open in X. The going-down property asserts that if $\mathfrak{p} \in E$ and $\mathfrak{p}' \subseteq \mathfrak{p}$, then $\mathfrak{p}' \in E$: in other words, that if X_0 is an irreducible closed subset of X and X_0 meets E, then $E \cap X_0$ is dense in X_0. By Exercises 20 and 22, E is open in X.]

25. Let A be Noetherian, $f: A \to B$ of finite type and *flat* (i.e., B is flat as an A-module). Then $f^*: \operatorname{Spec}(B) \to \operatorname{Spec}(A)$ is an open mapping. [Exercise 24 and Chapter 5, Exercise 11.]

Grothendieck groups

26. Let A be a Noetherian ring and let $F(A)$ denote the set of all isomorphism classes of finitely generated A-modules. Let C be the free abelian group generated by $F(A)$. With each short exact sequence $0 \to M' \to M \to M'' \to 0$ of finitely generated A-modules we associate the element $(M') - (M) + (M'')$ of C, where (M) is the isomorphism class of M, etc. Let D be the subgroup of C generated by these elements, for all short exact sequences. The quotient group C/D is called the *Grothendieck group* of A, and is denoted by $K(A)$. If M is a finitely generated A-module, let $\gamma(M)$, or $\gamma_A(M)$, denote the image of (M) in $K(A)$.
   i) Show that $K(A)$ has the following universal property: for each additive function λ on the class of finitely generated A-modules, with values in an abelian group G, there exists a unique homomorphism $\lambda_0: K(A) \to G$ such that $\lambda(M) = \lambda_0(\gamma(M))$ for all M.
   ii) Show that $K(A)$ is generated by the elements $\gamma(A/\mathfrak{p})$, where $\mathfrak{p}$ is a prime ideal of A. [Use Exercise 18.]
   iii) If A is a field, or more generally if A is a principal ideal domain, then $K(A) \cong \mathbf{Z}$.
   iv) Let $f: A \to B$ be a *finite* ring homomorphism. Show that restriction of scalars gives rise to a homomorphism $f_!: K(B) \to K(A)$ such that $f_!(\gamma_B(N)) = \gamma_A(N)$ for a B-module N. If $g: B \to C$ is another finite ring homomorphism, show that $(g \circ f)_! = f_! \circ g_!$.

27. Let A be a Noetherian ring and let $F_1(A)$ be the set of all isomorphism classes of finitely generated *flat* A-modules. Repeating the construction of Exercise 26 we obtain a group $K_1(A)$. Let $\gamma_1(M)$ denote the image of (M) in $K_1(A)$.
   i) Show that tensor product of modules over A induces a commutative ring structure on $K_1(A)$, such that $\gamma_1(M) \cdot \gamma_1(N) = \gamma_1(M \otimes N)$. The identity element of this ring is $\gamma_1(A)$.
   ii) Show that tensor product induces a $K_1(A)$-module structure on the group $K(A)$, such that $\gamma_1(M) \cdot \gamma(N) = \gamma(M \otimes N)$.
   iii) If A is a (Noetherian) local ring, then $K_1(A) \cong \mathbf{Z}$.
   iv) Let $f: A \to B$ be a ring homomorphism, B being Noetherian. Show that extension of scalars gives rise to a ring homomorphism $f^!: K_1(A) \to K_1(B)$ such that $f^!(\gamma_1(M)) = \gamma_1(B \otimes_A M)$. [If M is flat and finitely generated over A, then $B \otimes_A M$ is flat and finitely generated over B.] If $g: B \to C$ is another ring homomorphism (with C Noetherian), then $(f \circ g)^! = f^! \circ g^!$.
   v) If $f: A \to B$ is a finite ring homomorphism then
   $$f_!(f^!(x)y) = xf_!(y)$$
   for $x \in K_1(A)$, $y \in K(B)$. In other words, regarding $K(B)$ as a $K_1(A)$-module by restriction of scalars, the homomorphism $f_!$ is a $K_1(A)$-module homomorphism.

Remark. Since $F_1(A)$ is a subset of $F(A)$ we have a group homomorphism $\epsilon: K_1(A) \to K(A)$, given by $\epsilon(\gamma_1(M)) = \gamma(M)$. If the ring A is finite-dimensional and *regular*, i.e., if all its local rings $A_\mathfrak{p}$ are regular (Chapter 11) it can be shown that ϵ is an isomorphism.

8

Artin Rings

An *Artin ring* is one which satisfies the d.c.c. (or equivalently the minimal condition) on ideals.

The apparent symmetry with Noetherian rings is however misleading. In fact we will show that an Artin ring is necessarily Noetherian and of a very special kind. In a sense an Artin ring is the simplest kind of ring after a field, and we study them not because of their generality but because of their simplicity.

Proposition 8.1. *In an Artin ring A every prime ideal is maximal.*

Proof. Let $\mathfrak{p}$ be a prime ideal of A. Then $B = A/\mathfrak{p}$ is an Artinian integral domain. Let $x \in B$, $x \neq 0$. By the d.c.c. we have $(x^n) = (x^{n+1})$ for some n, hence $x^n = x^{n+1}y$ for some $y \in B$. Since B is an integral domain and $x \neq 0$, it follows that we may cancel x^n, hence $xy = 1$. Hence x has an inverse in B, and therefore B is a field, so that $\mathfrak{p}$ is a maximal ideal. ∎

Corollary 8.2. *In an Artin ring the nilradical is equal to the Jacobson radical.* ∎

Proposition 8.3. *An Artin ring has only a finite number of maximal ideals.*

Proof. Consider the set of all finite intersections $\mathfrak{m}_1 \cap \cdots \cap \mathfrak{m}_r$, where the $\mathfrak{m}_i$ are maximal ideals. This set has a minimal element, say $\mathfrak{m}_1 \cap \cdots \cap \mathfrak{m}_n$; hence for any maximal ideal $\mathfrak{m}$ we have $\mathfrak{m} \cap \mathfrak{m}_1 \cap \cdots \cap \mathfrak{m}_n = \mathfrak{m}_1 \cap \cdots \cap \mathfrak{m}_n$, and therefore $\mathfrak{m} \supseteq \mathfrak{m}_1 \cap \cdots \cap \mathfrak{m}_n$. By (1.11) $\mathfrak{m} \supseteq \mathfrak{m}_i$ for some i, hence $\mathfrak{m} = \mathfrak{m}_i$ since $\mathfrak{m}_i$ is maximal. ∎

Proposition 8.4. *In an Artin ring the nilradical $\mathfrak{N}$ is nilpotent.*

Proof. By d.c.c. we have $\mathfrak{N}^k = \mathfrak{N}^{k+1} = \cdots = \mathfrak{a}$ say, for some $k > 0$. Suppose $\mathfrak{a} \neq 0$, and let Σ denote the set of all ideals $\mathfrak{b}$ such that $\mathfrak{a}\mathfrak{b} \neq 0$. Then Σ is not empty, since $\mathfrak{a} \in \Sigma$. Let $\mathfrak{c}$ be a minimal element of Σ; then there exists $x \in \mathfrak{c}$ such that $x\mathfrak{a} \neq 0$; we have $(x) \subseteq \mathfrak{c}$, hence $(x) = \mathfrak{c}$ by the minimality of $\mathfrak{c}$. But $(x\mathfrak{a})\mathfrak{a} = x\mathfrak{a}^2 = x\mathfrak{a} \neq 0$, and $x\mathfrak{a} \subseteq (x)$, hence $x\mathfrak{a} = (x)$ (again by minimality). Hence $x = xy$ for some $y \in \mathfrak{a}$, and therefore $x = xy = xy^2 = \cdots = xy^n = \cdots$. But $y \in \mathfrak{a} = \mathfrak{N}^k \supseteq \mathfrak{N}$, hence y is nilpotent and therefore $x = xy^n = 0$. This contradicts the choice of x, therefore $\mathfrak{a} = 0$. ∎

By a *chain* of prime ideals of a ring A we mean a finite strictly increasing sequence $\mathfrak{p}_0 \subset \mathfrak{p}_1 \subset \cdots \subset \mathfrak{p}_n$; the *length* of the chain is n. We define the *dimension* of A to be the supremum of the lengths of all chains of prime ideals in A: it is an integer $\geqslant 0$, or $+\infty$ (assuming $A \neq 0$). A field has dimension 0; the ring $\mathbf{Z}$ has dimension 1.

Theorem 8.5. *A ring A is Artin* $\Leftrightarrow$ *A is Noetherian and* $\dim A = 0$.

Proof. $\Rightarrow$: By (8.1) we have $\dim A = 0$. Let $\mathfrak{m}_i$ $(1 \leqslant i \leqslant n)$ be the distinct maximal ideals of A (8.3). Then $\prod_{i=1}^{n} \mathfrak{m}_i^k \subseteq (\bigcap_{i=1}^{n} \mathfrak{m}_i)^k = \mathfrak{N}^k = 0$. Hence by (6.11) A is Noetherian.

$\Leftarrow$: Since the zero ideal has a primary decomposition (7.13), A has only a finite number of minimal prime ideals, and these are all maximal since $\dim A = 0$. Hence $\mathfrak{N} = \bigcap_{i=1}^{n} \mathfrak{m}_i$ say; we have $\mathfrak{N}^k = 0$ by (7.15), hence $\prod_{i=1}^{n} \mathfrak{m}_i^k = 0$ as in the previous part of the proof. Hence by (6.11) A is an Artin ring. ∎

If A is an Artin local ring with maximal ideal $\mathfrak{m}$, then $\mathfrak{m}$ is the only prime ideal of A and therefore $\mathfrak{m}$ is the nilradical of A. Hence every element of $\mathfrak{m}$ is nilpotent, and $\mathfrak{m}$ itself is nilpotent. Every element of A is either a unit or is nilpotent. An example of such a ring is $\mathbf{Z}/(p^n)$, where p is prime and $n \geqslant 1$.

Proposition 8.6. *Let A be a Noetherian local ring,* $\mathfrak{m}$ *its maximal ideal. Then exactly one of the following two statements is true:*

i) $\mathfrak{m}^n \neq \mathfrak{m}^{n+1}$ *for all n;*

ii) $\mathfrak{m}^n = 0$ *for some n, in which case A is an Artin local ring.*

Proof. Suppose $\mathfrak{m}^n = \mathfrak{m}^{n+1}$ for some n. By Nakayama's lemma (2.6) we have $\mathfrak{m}^n = 0$. Let $\mathfrak{p}$ be any prime ideal of A. Then $\mathfrak{m}^n \subseteq \mathfrak{p}$, hence (taking radicals) $\mathfrak{m} = \mathfrak{p}$. Hence $\mathfrak{m}$ is the only prime ideal of A and therefore A is Artinian. ∎

Theorem 8.7. (structure theorem for Artin rings). *An Artin ring A is uniquely (up to isomorphism) a finite direct product of Artin local rings.*

Proof. Let $\mathfrak{m}_i$ $(1 \leqslant i \leqslant n)$ be the distinct maximal ideals of A. From the proof of (8.5) we have $\prod_{i=1}^{n} \mathfrak{m}_i^k = 0$ for some $k > 0$. By (1.16) the ideals $\mathfrak{m}_i^k$ are coprime in pairs, hence $\bigcap \mathfrak{m}_i^k = \prod \mathfrak{m}_i^k$ by (1.10). Consequently by (1.10) again the natural mapping $A \to \prod_{i=1}^{n} (A/\mathfrak{m}_i^k)$ is an isomorphism. Each $A/\mathfrak{m}_i^k$ is an Artin local ring, hence A is a direct product of Artin local rings.

Conversely, suppose $A \cong \prod_{i=1}^{m} A_i$, where the A_i are Artin local rings. Then for each i we have a natural surjective homomorphism (projection on the ith factor) $\phi_i: A \to A_i$. Let $\mathfrak{a}_i = \operatorname{Ker}(\phi_i)$. By (1.10) the $\mathfrak{a}_i$ are pairwise coprime, and $\bigcap \mathfrak{a}_i = 0$. Let $\mathfrak{q}_i$ be the unique prime ideal of A_i, and let $\mathfrak{p}_i$ be its contraction $\phi_i^{-1}(\mathfrak{q}_i)$. The ideal $\mathfrak{p}_i$ is prime and therefore maximal by (8.1). Since $\mathfrak{q}_i$ is nilpotent it follows that $\mathfrak{a}_i$ is $\mathfrak{p}_i$-primary, and hence $\bigcap \mathfrak{a}_i = (0)$ is a primary decomposition of the zero ideal in A. Since the $\mathfrak{a}_i$ are pairwise coprime, so are the $\mathfrak{p}_i$, and they are therefore isolated prime ideals of (0). Hence all the primary components $\mathfrak{a}_i$ are isolated, and therefore uniquely determined by A, by the 2nd uniqueness theorem (4.11). Hence the rings $A_i \cong A/\mathfrak{a}_i$ are uniquely determined by A. ∎

Example. A ring with only one prime ideal need not be Noetherian (and hence not an Artin ring). Let $A = k[x_1, x_2, \ldots]$ be the polynomial ring in a countably infinite set of indeterminates x_n over a field k, and let $\mathfrak{a}$ be the ideal $(x_1, x_2^2, \ldots, x_n^n, \ldots)$. The ring $B = A/\mathfrak{a}$ has only one prime ideal (namely the image of $(x_1, x_2, \ldots, x_n, \ldots)$), hence B is a local ring of dimension 0. But B is not Noetherian, for it is not difficult to see that its prime ideal is not finitely generated.

If A is a local ring, $\mathfrak{m}$ its maximal ideal, $k = A/\mathfrak{m}$ its residue field, the A-module $\mathfrak{m}/\mathfrak{m}^2$ is annihilated by $\mathfrak{m}$ and therefore has the structure of a k-vector space. If $\mathfrak{m}$ is finitely generated (e.g., if A is Noetherian), the images in $\mathfrak{m}/\mathfrak{m}^2$ of a set of generators of $\mathfrak{m}$ will span $\mathfrak{m}/\mathfrak{m}^2$ as a vector space, and therefore $\dim_k(\mathfrak{m}/\mathfrak{m}^2)$ is finite. (See (2.8).)

Proposition 8.8. *Let A be an Artin local ring. Then the following are equivalent:*

i) *every ideal in A is principal;*

ii) *the maximal ideal $\mathfrak{m}$ is principal;*

iii) $\dim_k(\mathfrak{m}/\mathfrak{m}^2) \leqslant 1$.

Proof. i) $\Rightarrow$ ii) $\Rightarrow$ iii) is clear.

iii) $\Rightarrow$ i): If $\dim_k(\mathfrak{m}/\mathfrak{m}^2) = 0$, then $\mathfrak{m} = \mathfrak{m}^2$, hence $\mathfrak{m} = 0$ by Nakayama's lemma (2.6), and therefore A is a field and there is nothing to prove.

If $\dim_k(\mathfrak{m}/\mathfrak{m}^2) = 1$, then $\mathfrak{m}$ is a principal ideal by (2.8) (take $M = \mathfrak{m}$ there), say $\mathfrak{m} = (x)$. Let $\mathfrak{a}$ be an ideal of A, other than (0) or (1). We have $\mathfrak{m} = \mathfrak{N}$, hence $\mathfrak{m}$ is nilpotent by (8.4) and therefore there exists an integer r such that $\mathfrak{a} \subseteq \mathfrak{m}^r$, $\mathfrak{a} \nsubseteq \mathfrak{m}^{r+1}$; hence there exists $y \in \mathfrak{a}$ such that $y = ax^r$, $y \notin (x^{r+1})$; consequently $a \notin (x)$ and a is a unit in A. Hence $x^r \in \mathfrak{a}$, therefore $\mathfrak{m}^r = (x^r) \subseteq \mathfrak{a}$ and hence $\mathfrak{a} = \mathfrak{m}^r = (x^r)$. Hence $\mathfrak{a}$ is principal. ∎

Example. The rings $\mathbf{Z}/(p^n)$ (p prime), $k[x]/(f^n)$ (f irreducible) satisfy the conditions of (8.7). On the other hand, the Artin local ring $k[x^2, x^3]/(x^4)$ does not: here $\mathfrak{m}$ is generated by x^2 and x^3 (mod x^4), so that $\mathfrak{m}^2 = 0$ and $\dim(\mathfrak{m}/\mathfrak{m}^2) = 2$.

EXERCISES

1. Let $\mathfrak{q}_1 \cap \cdots \cap \mathfrak{q}_n = 0$ be a minimal primary decomposition of the zero ideal in a Noetherian ring, and let $\mathfrak{q}_i$ be $\mathfrak{p}_i$-primary. Let $\mathfrak{p}_i^{(r)}$ be the rth *symbolic power* of $\mathfrak{p}_i$ (Chapter 4, Exercise 13). Show that for each $i = 1, \ldots, n$ there exists an integer r_i such that $\mathfrak{p}_i^{(r_i)} \subseteq \mathfrak{q}_i$.

   Suppose $\mathfrak{q}_i$ is an isolated primary component. Then $A_{\mathfrak{p}_i}$ is an Artin local ring, hence if $\mathfrak{m}_i$ is its maximal ideal we have $\mathfrak{m}_i^r = 0$ for all sufficiently large r, hence $\mathfrak{q}_i = \mathfrak{p}_i^{(r)}$ for all large r.

If $\mathfrak{q}_i$ is an embedded primary component, then $A_{\mathfrak{p}_i}$ is *not* Artinian, hence the powers $\mathfrak{m}_i^r$ are all distinct, and so the $\mathfrak{p}_i^{(r)}$ are all distinct. Hence in the given primary decomposition we can replace $\mathfrak{q}_i$ by any of the infinite set of $\mathfrak{p}_i$-primary ideals $\mathfrak{p}_i^{(r)}$ where $r \geqslant r_i$, and so there are infinitely many minimal primary decompositions of 0 which differ only in the $\mathfrak{p}_i$-component.

2. Let A be a Noetherian ring. Prove that the following are equivalent:
   i) A is Artinian;
   ii) Spec (A) is discrete and finite;
   iii) Spec (A) is discrete.

3. Let k be a field and A a finitely generated k-algebra. Prove that the following are equivalent:
   i) A is Artinian;
   ii) A is a finite k-algebra.
   [To prove that i) $\Rightarrow$ ii), use (8.7) to reduce to the case where A is an Artin local ring. By the Nullstellensatz, the residue field of A is a finite extension of k. Now use the fact that A is of finite length as an A-module. To prove ii) $\Rightarrow$ i), observe that the ideals of A are k-vector subspaces and therefore satisfy d.c.c.]

4. Let $f: A \rightarrow B$ be a ring homomorphism of finite type. Consider the following statements:
   i) f is finite;
   ii) the fibres of f^* are discrete subspaces of Spec (B);
   iii) for each prime ideal $\mathfrak{p}$ of A, the ring $B \otimes_A k(\mathfrak{p})$ is a finite $k(\mathfrak{p})$-algebra ($k(\mathfrak{p})$ is the residue field of $A_{\mathfrak{p}}$);
   iv) the fibres of f^* are finite.
   Prove that i) $\Rightarrow$ ii) $\Leftrightarrow$ iii) $\Rightarrow$ iv). [Use Exercises 2 and 3.]
   If f is integral and the fibres of f^* are finite, is f necessarily finite?

5. In Chapter 5, Exercise 16, show that X is a finite covering of L (i.e., the number of points of X lying over a given point of L is finite and bounded).

6. Let A be a Noetherian ring and $\mathfrak{q}$ a $\mathfrak{p}$-primary ideal in A. Consider chains of primary ideals from $\mathfrak{q}$ to $\mathfrak{p}$. Show that all such chains are of finite bounded length, and that all maximal chains have the same length.

9

Discrete Valuation Rings and Dedekind Domains

As we have indicated before, algebraic number theory is one of the historical sources of commutative algebra. In this chapter we specialize down to the case of interest in number theory, namely to Dedekind domains. We deduce the unique factorization of ideals in Dedekind domains from the general primary decomposition theorems. Although a direct approach is of course possible one obtains more insight our way into the precise context of number theory in commutative algebra. Another important class of Dedekind domains occurs in connection with non-singular algebraic curves. In fact the geometrical picture of the Dedekind condition is: non-singular of dimension one.

The last chapter dealt with Noetherian rings of dimension 0. Here we start by considering the next simplest case, namely Noetherian *integral domains* of dimension one: i.e., Noetherian domains in which every non-zero prime ideal is maximal. The first result is that in such a ring we have a unique factorization theorem for ideals:

Proposition 9.1. *Let A be a Noetherian domain of dimension 1. Then every non-zero ideal $\mathfrak{a}$ in A can be uniquely expressed as a product of primary ideals whose radicals are all distinct.*

Proof. Since A is Noetherian, $\mathfrak{a}$ has a minimal primary decomposition $\mathfrak{a} = \bigcap_{i=1}^{n} \mathfrak{q}_i$ by (7.13), where each $\mathfrak{q}_i$ is say $\mathfrak{p}_i$-primary. Since $\dim A = 1$ and A is an integral domain, each non-zero prime ideal of A is maximal, hence the $\mathfrak{p}_i$ are distinct maximal ideals (since $\mathfrak{p}_i \supseteq \mathfrak{q}_i \supseteq \mathfrak{a} \neq 0$), and are therefore pairwise coprime. Hence by (1.16) the $\mathfrak{q}_i$ are pairwise coprime and therefore by (1.10) we have $\prod \mathfrak{q}_i = \bigcap \mathfrak{q}_i$. Hence $\mathfrak{a} = \prod \mathfrak{q}_i$.

Conversely, if $\mathfrak{a} = \prod \mathfrak{q}_i$, the same argument shows that $\mathfrak{a} = \bigcap \mathfrak{q}_i$; this is a minimal primary decomposition of $\mathfrak{a}$, in which each $\mathfrak{q}_i$ is an isolated primary component, and is therefore unique by (4.11). ∎

Let A be a Noetherian domain of dimension one in which every primary ideal is a prime power. By (9.1), in such a ring we shall have unique factorization of non-zero ideals into products of prime ideals. If we localize A with respect to a non-zero prime ideal $\mathfrak{p}$ we get a local ring $A_\mathfrak{p}$ satisfying the same conditions as A, and therefore in $A_{\mathfrak{p}}$ every non-zero ideal is a power of the maximal ideal. Such local rings can be characterized in other ways.

DISCRETE VALUATION RINGS

Let K be a field. A *discrete valuation* on K is a mapping v of K^* onto $\mathbf{Z}$ (where $K^* = K - \{0\}$ is the multiplicative group of K) such that

1) $v(xy) = v(x) + v(y)$, i.e., v is a *homomorphism*;
2) $v(x + y) \geqslant \min(v(x), v(y))$.

The set consisting of 0 and all $x \in K^*$ such that $v(x) \geqslant 0$ is a ring, called the *valuation ring* of v. It is a valuation ring of the field K. It is sometimes convenient to extend v to the whole of K by putting $v(0) = +\infty$.

Examples. The two standard examples are:

1) $K = \mathbf{Q}$. Take a fixed prime p, then any non zero $x \in \mathbf{Q}$ can be written uniquely in the form $p^a y$, where $a \in \mathbf{Z}$ and both numerator and denominator of y are prime to p. Define $v_p(x)$ to be a. The valuation ring of v_p is the local ring $\mathbf{Z}_{(p)}$.

2) $K = k(x)$, where k is a field and x an indeterminate. Take a fixed irreducible polynomial $f \in k[x]$ and define v_f just as in 1). The valuation ring of v_f is then the local ring of $k[x]$ with respect to the prime ideal (f).

An integral domain A is a *discrete valuation ring* if there is a discrete valuation v of its field of fractions K such that A is the valuation ring of v. By (5.18), A is a local ring, and its maximal ideal $\mathfrak{m}$ is the set of all $x \in K$ such that $v(x) > 0$.

If two elements x, y of A have the same value, that is if $v(x) = v(y)$, then $v(xy^{-1}) = 0$ and therefore $u = xy^{-1}$ is a unit in A. Hence $(x) = (y)$.

If $\mathfrak{a} \neq 0$ is an ideal in A, there is a least integer k such that $v(x) = k$ for some $x \in \mathfrak{a}$. It follows that $\mathfrak{a}$ contains every $y \in A$ with $v(y) \geqslant k$, and therefore the only ideals $\neq 0$ in A are the ideals $\mathfrak{m}_k = \{y \in A : v(y) \geqslant k\}$. These form a single chain $\mathfrak{m} \supset \mathfrak{m}_2 \supset \mathfrak{m}_3 \supset \cdots$, and therefore A is Noetherian.

Moreover, since $v: K^* \to Z$ is surjective, there exists $x \in \mathfrak{m}$ such that $v(x) = 1$, and then $\mathfrak{m} = (x)$, and $\mathfrak{m}_k = (x^k)$ $(k \geqslant 1)$. Hence $\mathfrak{m}$ is the only non-zero prime ideal of A, and A is thus a Noetherian local domain of dimension one in which every non-zero ideal is a power of the maximal ideal. In fact many of these properties are characteristic of discrete valuation rings.

Proposition 9.2. *Let A be a Noetherian local domain of dimension one, $\mathfrak{m}$ its maximal ideal, $k = A/\mathfrak{m}$ its residue field. Then the following are equivalent:*

i) *A is a discrete valuation ring;*

ii) *A is integrally closed;*

iii) *$\mathfrak{m}$ is a principal ideal;*

iv) $\dim_k(\mathfrak{m}/\mathfrak{m}^2) = 1$;

v) *Every non-zero ideal is a power of* $\mathfrak{m}$;

vi) *There exists* $x \in A$ *such that every non-zero ideal is of the form* (x^k), $k \geq 0$.

Proof. Before we start going the rounds, we make two remarks:

(A) If $\mathfrak{a}$ is an ideal $\neq 0, (1)$, then $\mathfrak{a}$ is $\mathfrak{m}$-primary and $\mathfrak{a} \supseteq \mathfrak{m}^n$ for some n. For $r(\mathfrak{a}) = \mathfrak{m}$, since $\mathfrak{m}$ is the only non-zero prime ideal; now use (7.16).

(B) $\mathfrak{m}^n \neq \mathfrak{m}^{n+1}$ for all $n \geq 0$. This follows from (8.6).

i) $\Rightarrow$ ii) by (5.18).

ii) $\Rightarrow$ iii). Let $a \in \mathfrak{m}$ and $a \neq 0$. By remark (A) there exists an integer n such that $\mathfrak{m}^n \subseteq (a)$, $\mathfrak{m}^{n-1} \nsubseteq (a)$. Choose $b \in \mathfrak{m}^{n-1}$ and $b \notin (a)$, and let $x = a/b \in K$, the field of fractions of A. We have $x^{-1} \notin A$ (since $b \notin (a)$), hence x^{-1} is not integral over A, and therefore by (5.1) we have $x^{-1}\mathfrak{m} \nsubseteq \mathfrak{m}$ (for if $x^{-1}\mathfrak{m} \subseteq \mathfrak{m}$, $\mathfrak{m}$ would be a faithful $A[x^{-1}]$-module, finitely generated as an A-module). But $x^{-1}\mathfrak{m} \subseteq A$ by construction of x, hence $x^{-1}\mathfrak{m} = A$ and therefore $\mathfrak{m} = Ax = (x)$.

iii) $\Rightarrow$ iv). By (2.8) we have $\dim_k(\mathfrak{m}/\mathfrak{m}^2) \leq 1$, and by remark (B) $\mathfrak{m}/\mathfrak{m}^2 \neq 0$.

iv) $\Rightarrow$ v). Let $\mathfrak{a}$ be an ideal $\neq (0), (1)$. By remark (A) we have $\mathfrak{a} \supseteq \mathfrak{m}^n$ for some n; from (8.8) (applied to $A/\mathfrak{m}^n$) it follows that $\mathfrak{a}$ is a power of $\mathfrak{m}$.

v) $\Rightarrow$ vi). By remark (B), $\mathfrak{m} \neq \mathfrak{m}^2$, hence there exists $x \in \mathfrak{m}$, $x \notin \mathfrak{m}^2$. But $(x) = \mathfrak{m}^r$ by hypothesis, hence $r = 1$, $(x) = \mathfrak{m}$, $(x^k) = \mathfrak{m}^k$.

vi) $\Rightarrow$ i). Clearly $(x) = \mathfrak{m}$, hence $(x^k) \neq (x^{k+1})$ by remark (B). Hence if a is any non-zero element of A, we have $(a) = (x^k)$ for exactly one value of k. Define $v(a) = k$ and extend v to K^* by defining $v(ab^{-1}) = v(a) - v(b)$. Check that v is well-defined and is a discrete valuation, and that A is the valuation ring of v. ∎

DEDEKIND DOMAINS

Theorem 9.3. *Let A be a Noetherian domain of dimension one. Then the following are equivalent:*

i) *A is integrally closed;*

ii) *Every primary ideal in A is a prime power;*

iii) *Every local ring $A_\mathfrak{p}$ ($\mathfrak{p} \neq 0$) is a discrete valuation ring.*

Proof. i) $\Leftrightarrow$ iii) by (9.2) and (5.13).

ii) $\Leftrightarrow$ iii). Use (9.2) and the fact that primary ideals and powers of ideals behave well under localization: (4.8), (3.11). ∎

A ring satisfying the conditions of (9.3) is called a *Dedekind domain.*

Corollary 9.4. *In a Dedekind domain every non-zero ideal has a unique factorization as a product of prime ideals.*

Proof. (9.1) and (9.3). ∎

Examples. 1) Any principal ideal domain A. For A is Noetherian (since every ideal is finitely generated) and of dimension one (Example 3 after (1.6)). Also every local ring $A_{\mathfrak{p}}(\mathfrak{p} \neq 0)$ is a principal ideal domain, hence by (9.2) a discrete valuation ring; hence A is a Dedekind domain by (9.3).

2) Let K be an algebraic number field (a finite algebraic extension of $\mathbf{Q}$). Its *ring of integers* A is the integral closure of $\mathbf{Z}$ in K. (For example, if $K = \mathbf{Q}(i)$, then $A = \mathbf{Z}[i]$, the ring of Gaussian integers.) Then A is a Dedekind domain:

Theorem 9.5. *The ring of integers in an algebraic number field K is a Dedekind domain.*

Proof. K is a separable extension of $\mathbf{Q}$ (because the characteristic is zero), hence by (5.17) there is a basis $v_1, \ldots, v_n$ of K over $\mathbf{Q}$ such that $A \subseteq \sum \mathbf{Z}v_j$. Hence A is finitely generated as a $\mathbf{Z}$-module and therefore Noetherian. Also A is integrally closed by (5.5). To complete the proof we must show that every non-zero prime ideal $\mathfrak{p}$ of A is maximal, and this follows from (5.8) and (5.9): (5.9) shows that $\mathfrak{p} \cap \mathbf{Z} \neq 0$, hence $\mathfrak{p} \cap \mathbf{Z}$ is a maximal ideal of $\mathbf{Z}$ and therefore $\mathfrak{p}$ is maximal in A by (5.8). ∎

Remark. The unique factorization theorem (9.4) was originally proved for rings of integers in algebraic number fields. The uniqueness theorems of Chapter 4 may be regarded as generalizations of this result: prime powers have to be replaced by primary ideals, and products by intersections.

FRACTIONAL IDEALS

Let A be an integral domain, K its field of fractions. An A-submodule M of K is a *fractional ideal* of A if $xM \subseteq A$ for some $x \neq 0$ in A. In particular, the "ordinary" ideals (now called *integral* ideals) are fractional ideals (take $x = 1$). Any element $u \in K$ generates a fractional ideal, denoted by (u) or Au, and called *principal*. If M is a fractional ideal, the set of all $x \in K$ such that $x\,M \subseteq A$ is denoted by $(A:M)$.

Every finitely generated A-submodule M of K is a fractional ideal. For if M is generated by $x_1, \ldots, x_n \in K$, we can write $x_i = y_i z^{-1}$ $(1 \leqslant i \leqslant n)$ where y_i and z are in A, and then $zM \subseteq A$. Conversely, if A is Noetherian, every fractional ideal is finitely generated, for it is of the form $x^{-1}\mathfrak{a}$ for some integral ideal $\mathfrak{a}$.

An A-submodule M of K is an *invertible ideal* if there exists a submodule N of K such that $MN = A$. The module N is then unique and equal to $(A:M)$, for we have $N \subseteq (A:M) = (A:M)MN \subseteq AN = N$. It follows that M is finitely generated, and therefore a fractional ideal: for since $M\cdot(A:M) = A$ there exist $x_i \in M$ and $y_i \in (A:M)$ $(1 \leqslant i \leqslant n)$ such that $\sum x_i y_i = 1$, and hence for any $x \in M$ we have $x = \sum (y_i x)x_i$: each $y_i x \in A$, so that M is generated by $x_1, \ldots, x_n$.

Clearly every non-zero principal fractional ideal (u) is invertible, its inverse being (u^{-1}). The invertible ideals form a group with respect to multiplication, whose identity element is $A = (1)$.

Invertibility is a *local* property:

Proposition 9.6. *For a fractional ideal M, the following are equivalent:*

i) *M is invertible;*

ii) *M is finitely generated and, for each prime ideal $\mathfrak{p}$, $M_\mathfrak{p}$ is invertible:*

iii) *M is finitely generated and, for each maximal ideal $\mathfrak{m}$, $M_\mathfrak{m}$ is invertible.*

Proof. i) $\Rightarrow$ ii): $A_\mathfrak{p} = (M\cdot(A:M))_\mathfrak{p} = M_\mathfrak{p}\cdot(A_\mathfrak{p}:M_\mathfrak{p})$ by (3.11) and (3.15) (for M is finitely generated, because invertible).

ii) $\Rightarrow$ iii) as usual.

iii $\Rightarrow$ i): Let $\mathfrak{a} = M\cdot(A:M)$, which is an integral ideal. For each maximal ideal $\mathfrak{m}$ we have $\mathfrak{a}_\mathfrak{m} = M_\mathfrak{m}\cdot(A_\mathfrak{m}:M_\mathfrak{m})$ (by (3.11) and (3.15)) $= A_\mathfrak{m}$ because $M_\mathfrak{m}$ is invertible. Hence $\mathfrak{a} \nsubseteq \mathfrak{m}$. Consequently $\mathfrak{a} = A$ and therefore M is invertible. ■

Proposition 9.7. *Let A be a local domain. Then A is a discrete valuation ring $\Leftrightarrow$ every non-zero fractional ideal of A is invertible.*

Proof. $\Rightarrow$. Let x be a generator of the maximal ideal $\mathfrak{m}$ of A, and let $M \neq 0$ be a fractional ideal. Then there exists $y \in A$ such that $yM \subseteq A$: thus yM is an integral ideal, say (x^r), and therefore $M = (x^{r-s})$ where $s = v(y)$.

$\Leftarrow$: Every non-zero integral ideal is invertible and therefore finitely generated, so that A is Noetherian. It is therefore enough to prove that every non-zero integral ideal is a power of $\mathfrak{m}$. Suppose this is false; let Σ be the set of non-zero ideals which are not powers of $\mathfrak{m}$, and let $\mathfrak{a}$ be a maximal element of Σ. Then $\mathfrak{a} \neq \mathfrak{m}$, hence $\mathfrak{a} \subset \mathfrak{m}$; hence $\mathfrak{m}^{-1}\mathfrak{a} \subset \mathfrak{m}^{-1}\mathfrak{m} = A$ is a proper (integral) ideal, and $\mathfrak{m}^{-1}\mathfrak{a} \supseteq \mathfrak{a}$. If $\mathfrak{m}^{-1}\mathfrak{a} = \mathfrak{a}$, then $\mathfrak{a} = \mathfrak{m}\mathfrak{a}$ and therefore $\mathfrak{a} = 0$ by Nakayama's lemma (2.6); hence $\mathfrak{m}^{-1}\mathfrak{a} \supset \mathfrak{a}$ and hence $\mathfrak{m}^{-1}\mathfrak{a}$ is a power of $\mathfrak{m}$ (by the maximality of $\mathfrak{a}$). Hence $\mathfrak{a}$ is a power of $\mathfrak{m}$: contradiction. ■

The "global" counterpart of (9.7) is

Theorem 9.8. *Let A be an integral domain. Then A is a Dedekind domain $\Leftrightarrow$ every non-zero fractional ideal of A is invertible.*

Proof. $\Rightarrow$: Let $M \neq 0$ be a fractional ideal. Since A is Noetherian, M is finitely generated. For each prime ideal $\mathfrak{p} \neq 0$, $M_\mathfrak{p}$ is a fractional ideal $\neq 0$ of the discrete valuation ring $A_\mathfrak{p}$, hence is invertible by (9.7). Hence M is invertible, by (9.6).

$\Leftarrow$: Every non-zero integral ideal is invertible, hence finitely generated, hence A is Noetherian. We shall show that each $A_\mathfrak{p}$ ($\mathfrak{p} \neq 0$) is a discrete valuation ring. For this it is enough to show that each integral ideal $\neq 0$ in $A_\mathfrak{p}$ is invertible, and then use (9.7). Let $\mathfrak{b} \neq 0$ be an (integral) ideal in $A_\mathfrak{p}$, and let $\mathfrak{a} = \mathfrak{b}^c = \mathfrak{b} \cap A$. Then $\mathfrak{a}$ is invertible, hence $\mathfrak{b} = \mathfrak{a}_\mathfrak{p}$ is invertible by (9.7). ■

Corollary 9.9. If A is a Dedekind domain, the non-zero fractional ideals of A form a group with respect to multiplication. ∎

This group is called the *group of ideals* of A; we denote it by I. In this terminology (9.4) says that I is a free (abelian) group, generated by the non-zero prime ideals of A.

Let K^* denote the multiplicative group of the field of fractions K of A. Each $u \in K^*$ defines a fractional ideal (u), and the mapping $u \mapsto (u)$ is a homomorphism $\phi: K^* \to I$. The image P of ϕ is the group of *principal* fractional ideals: the quotient group $H = I/P$ is called the *ideal class group* of A. The kernel U of ϕ is the set of all $u \in K^*$ such that $(u) = (1)$, so that it is the *group of units* of A. We have an exact sequence

$$1 \to U \to K^* \to I \to H \to 1.$$

Remark. For the Dedekind domains that arise in number theory, there are classical theorems relating to the groups H and U. Let K be an algebraic number field and let A be its ring of integers, which is a Dedekind domain by (9.5). In this case:

1) H is a *finite* group. Its order h is the *class number* of the field K. The following are equivalent: (i) $h = 1$; (ii) $I = P$; (iii) A is a principal ideal domain; (iv) A is a unique factorization domain.

2) U is a *finitely-generated* abelian group. More precisely, we can specify the number of generators of U. First, the elements of finite order in U are just the roots of unity which lie in K, and they form a finite cyclic group W; U/W is torsion-free. The number of generators of U/W is given as follows: if $(K:\mathbf{Q}) = n$ there are n distinct embeddings $K \to \mathbf{C}$ (the field of complex numbers). Of these, say r_1 map K into $\mathbf{R}$, and the rest pair off (if α is one, then $\omega \circ \alpha$ is another, where ω is the automorphism of $\mathbf{C}$ defined by $\omega(z) = \bar{z}$) into say r_2 pairs: thus $r_1 + 2r_2 = n$. The number of generators of U/W is then $r_1 + r_2 - 1$.

The proofs of these results belong to algebraic number theory and not to commutative algebra: they require techniques of a different nature from those used in this book.

Examples. 1) $K = \mathbf{Q}(\sqrt{-1})$; $n = 2$, $r_1 = 0$, $r_2 = 1$, $r_1 + r_2 - 1 = 0$. The only units in $\mathbf{Z}[i] = A$ are the four roots of unity $\pm 1, \pm i$.

2) $K = \mathbf{Q}(\sqrt{2})$; $n = 2$, $r_1 = 2$, $r_2 = 0$, $r_1 + r_2 - 1 = 1$. $W = \{\pm 1\}$, and U/W is infinite cyclic. In fact the units in $A = \mathbf{Z}[\sqrt{2}]$ are $\pm(1 + \sqrt{2})^n$, where n is any rational integer.

EXERCISES

1. Let A be a Dedekind domain, S a multiplicatively closed subset of A. Show that $S^{-1}A$ is either a Dedekind domain or the field of fractions of A.
   Suppose that $S \neq A - \{0\}$, and let H, H' be the ideal class groups of A and $S^{-1}A$ respectively. Show that extension of ideals induces a surjective homomorphism $H \to H'$.

2. Let A be a Dedekind domain. If $f = a_0 + a_1x + \cdots + a_nx^n$ is a polynomial with coefficients in A, the *content* of f is the ideal $c(f) = (a_0, \ldots, a_n)$ in A. Prove *Gauss's lemma* that $c(fg) = c(f)c(g)$.
   [Localize at each maximal ideal.]

3. A valuation ring (other than a field) is Noetherian if and only if it is a discrete valuation ring.

4. Let A be a local domain which is not a field and in which the maximal ideal $\mathfrak{m}$ is principal and $\bigcap_{n=1}^{\infty} \mathfrak{m}^n = 0$. Prove that A is a discrete valuation ring.

5. Let M be a finitely-generated module over a Dedekind domain. Prove that M is flat $\Leftrightarrow$ M is torsion-free.
   [Use Chapter 3, Exercise 13 and Chapter 7, Exercise 16.]

6. Let M be a finitely-generated torsion module ($T(M) = M$) over a Dedekind domain A. Prove that M is uniquely representable as a finite direct sum of modules $A/\mathfrak{p}_i^{n_i}$, where $\mathfrak{p}_i$ are non-zero prime ideals of A. [For each $\mathfrak{p} \neq 0$, $M_\mathfrak{p}$ is a torsion $A_\mathfrak{p}$-module; use the structure theorem for modules over a principal ideal domain.]

7. Let A be a Dedekind domain and $\mathfrak{a} \neq 0$ an ideal in A. Show that every ideal in $A/\mathfrak{a}$ is principal.
   Deduce that every ideal in A can be generated by at most 2 elements.

8. Let $\mathfrak{a}, \mathfrak{b}, \mathfrak{c}$ be three ideals in a Dedekind domain. Prove that
$$\mathfrak{a} \cap (\mathfrak{b} + \mathfrak{c}) = (\mathfrak{a} \cap \mathfrak{b}) + (\mathfrak{a} \cap \mathfrak{c})$$
$$\mathfrak{a} + (\mathfrak{b} \cap \mathfrak{c}) = (\mathfrak{a} + \mathfrak{b}) \cap (\mathfrak{a} + \mathfrak{c}).$$
   [Localize.]

9. (Chinese Remainder Theorem). Let $\mathfrak{a}_1, \ldots, \mathfrak{a}_n$ be ideals and let $x_1, \ldots, x_n$ be elements in a Dedekind domain A. Then the system of congruences $x \equiv x_i \pmod{\mathfrak{a}_i}$ $(1 \leqslant i \leqslant n)$ has a solution x in A $\Leftrightarrow$ $x_i \equiv x_j \pmod{\mathfrak{a}_i + \mathfrak{a}_j}$ whenever $i \neq j$.
   [This is equivalent to saying that the sequence of A-modules
$$A \xrightarrow{\phi} \bigoplus_{i=1}^{n} A/\mathfrak{a}_i \xrightarrow{\psi} \bigoplus_{i<j} A/(\mathfrak{a}_i + \mathfrak{a}_j)$$
   is exact, where ϕ and ψ are defined as follows:
   $\phi(x) = (x + \mathfrak{a}_1, \ldots, x + \mathfrak{a}_n)$; $\psi(x_1 + \mathfrak{a}_1, \ldots, x_n + \mathfrak{a}_n)$ has (i, j)-component $x_i - x_j + \mathfrak{a}_i + \mathfrak{a}_j$. To show that this sequence is exact it is enough to show that it is exact when localized at any $\mathfrak{p} \neq 0$: in other words we may assume that A is a discrete valuation ring, and then it is easy.]

10

Completions

In classical algebraic geometry (i.e. over the field of complex numbers) we can use transcendental methods. This means that we regard a rational function as an analytic function (of one or more complex variables) and consider its power series expansion about a point. In abstract algebraic geometry the best we can do is to consider the corresponding formal power series. This is not so powerful as in the holomorphic case but it can be a very useful tool. The process of replacing polynomials by formal power series is an example of a general device known as *completion*. Another important instance of completion occurs in number theory in the formation of p-adic numbers. If p is a prime number in $\mathbf{Z}$ we can work in the various quotient rings $\mathbf{Z}/p^n\mathbf{Z}$: in other words, we can try and solve congruences modulo p^n for higher and higher values of n. This is analogous to the successive approximations given by the terms of a Taylor expansion and, just as it is convenient to introduce formal power series, so it is convenient to introduce the p-adic numbers, these being the limit in a certain sense of $\mathbf{Z}/p^n\mathbf{Z}$ as $n \to \infty$. In one respect, however, the p-adic numbers are more complicated than formal power series (in, say, one variable x). Whereas the polynomials of degree n are naturally embedded in the power series, the group $\mathbf{Z}/p^n\mathbf{Z}$ cannot be embedded in $\mathbf{Z}$. Although a p-adic integer can be thought of as a power series $\sum a_n p^n$ $(0 \leqslant a_n < p)$ this representation does not behave well under the ring operations.

In this chapter we shall describe the general process of "adic" completion—the prime p being replaced by a general ideal. It is most conveniently expressed in topological terms but the reader should beware of using the topology of the real numbers as an intuitive guide. Instead he should think of the power series topology in which a power series is "small" if it has only terms of high order. Alternatively he can think of the p-adic topology on $\mathbf{Z}$, in which an integer is "small" if it is divisible by a high power of p.

Completion, like localization, is a method of simplifying things by concentrating attention near a point (or prime). It is, however, a more drastic simplification than localization. For example, in algebraic geometry the local ring of a non-singular point on a variety of dimension n always has for its completion the ring of formal power series in n variables (this will essentially be proved in Chapter 11). On the other hand the local rings of two such points cannot be isomorphic unless the varieties on which they lie are birationally equivalent (this means that the fields of fractions of the two local rings are isomorphic).

Two of the important properties of localization are that it preserves exactness and the Noetherian property. The same is true for completion—when we restrict to finitely-generated modules—but the proofs are much harder and take up most of this chapter. Another important result is the theorem of Krull which identifies the part of a ring which is "killed" by completion. Roughly speaking, Krull's Theorem is the analogue of the fact that an analytic function is determined by the coefficients of its Taylor expansion. This analogy is clearest for a Noetherian local ring in which case the theorem just asserts that $\bigcap \mathfrak{m}^n = 0$ where $\mathfrak{m}$ is the maximal ideal. Both Krull's Theorem and the exactness of completion are easy consequences of the well-known "Artin–Rees Lemma", and we accord this lemma a central place in our treatment.

For the study of completions we shall find it necessary to introduce graded rings. The prototype of a graded ring is the ring of polynomials $k[x_1, \ldots, x_n]$, the grading being the usual one obtained by taking the degree of each variable to be 1. Just as ungraded rings are the foundation for affine algebraic geometry, so graded rings are the foundation for projective algebraic geometry. They are therefore of considerable geometric importance. The important construction of the associated graded ring $G_{\mathfrak{a}}(A)$ of an ideal $\mathfrak{a}$ of A, which we shall meet, has a very definite geometrical interpretation. For example, if A is the local ring of a point P on a variety V with $\mathfrak{a}$ as maximal ideal, then $G_{\mathfrak{a}}(A)$ corresponds to the projective tangent cone at P, i.e. all the lines through P which are tangent to V at P. This geometrical picture should help to explain the significance of $G_{\mathfrak{a}}(A)$ in connection with the properties of V near P and in particular in connection with the study of the completion $\hat{A}$.

TOPOLOGIES AND COMPLETIONS

Let G be a topological abelian group (written additively), not necessarily Hausdorff: thus G is both a topological space and an abelian group, and the two structures on G are compatible in the sense that the mappings $G \times G \to G$ and $G \to G$, defined by $(x, y) \mapsto x + y$ and $x \mapsto -x$ respectively, are continuous. If $\{0\}$ is closed in G, then the diagonal is closed in $G \times G$ (being the inverse image of $\{0\}$ under the mapping $(x, y) \mapsto x - y$) and so G is Hausdorff. If a is a fixed element of G the translation T_a defined by $T_a(x) = x + a$ is a homeomorphism of G onto G (for T_a is continuous, and its inverse is T_{-a}); hence if U is any neighborhood of 0 in G, then $U + a$ is a neighborhood of a in G, and conversely every neighborhood of a appears in this form. Thus the topology of G is uniquely determined by the neighborhoods of 0 in G.

Lemma 10.1. *Let H be the intersection of all neighborhoods of* 0 *in G. Then*
i) *H is a subgroup.*

ii) *H is the closure of* $\{0\}$.

iii) *G/H is Hausdorff.*

iv) *G is Hausdorff* $\Leftrightarrow H = 0$.

Proof. i) follows from the continuity of the group operations. For ii) we have:

$$\begin{aligned} x \in H &\Leftrightarrow 0 \in x - U \text{ for all neighborhoods } U \text{ of } 0 \\ &\Leftrightarrow x \in \overline{\{0\}}. \end{aligned}$$

ii) implies that the cosets of H are all closed; thus points are closed in G/H and so G/H is Hausdorff. Thus $H = 0 \Rightarrow G$ is Hausdorff, and the converse is trivial. ∎

Assume for simplicity that $0 \in G$ has a countable fundamental system of neighborhoods. Then the completion $\hat{G}$ of G may be defined in the usual way by means of Cauchy sequences. A *Cauchy sequence* in G is defined to be a sequence (x_ν) of elements of G such that, for any neighborhood U of 0, there exists an integer $s(U)$ with the property that

$$x_\mu - x_\nu \in U \text{ for all } \mu, \nu \geqslant s(U).$$

Two Cauchy sequences are *equivalent* if $x_\nu - y_\nu \to 0$ in G. The set of all equivalence classes of Cauchy sequences is denoted by $\hat{G}$. If (x_ν), (y_ν) are Cauchy sequences, so is $(x_\nu + y_\nu)$, and its class in $\hat{G}$ depends only on the classes of (x_ν) and (y_ν). Hence we have an addition in $\hat{G}$ with respect to which $\hat{G}$ is an abelian group. For each $x \in G$ the class of the constant sequence (x) is an element $\phi(x)$ of $\hat{G}$, and $\phi: G \to \hat{G}$ is a homomorphism of abelian groups. Note that ϕ is not in general injective. In fact we have

$$\operatorname{Ker} \phi = \bigcap U$$

where U runs through all neighborhoods of 0 in G, and so by (10.1) ϕ is injective if and only if G is Hausdorff.

If H is another abelian topological group and $f: G \to H$ a continuous homomorphism, then the image under f of a Cauchy sequence in G is a Cauchy sequence in H, and therefore f induces a homomorphism $\hat{f}: \hat{G} \to \hat{H}$, which is continuous. If we have $G \xrightarrow{f} H \xrightarrow{g} K$, then $\widehat{g \circ f} = \hat{g} \circ \hat{f}$.

So far we have been quite general and G could for instance have been the additive group of rationals with the usual topology, so that $\hat{G}$ would be the real numbers. Now, however, we restrict ourselves to the special kind of topologies occurring in commutative algebra, namely we assume that $0 \in G$ has a fundamental system of neighborhoods consisting of *subgroups*. Thus we have a sequence of subgroups

$$G = G_0 \supseteq G_1 \supseteq \cdots \supseteq G_n \supseteq \cdots$$

and $U \subseteq G$ is a neighborhood of 0 if and only if it contains some G_n. A typical example is the p-adic topology on $\mathbf{Z}$, in which $G_n = p^n\mathbf{Z}$. Note that in such topologies the subgroups G_n of G are both open and closed. In fact if $g \in G_n$

then $g + G_n$ is a neighborhood of g; since $g + G_n \subseteq G_n$ this shows G_n is open. Hence for any h the coset $h + G_n$ is open and therefore $\bigcup_{h \notin G_n} (h + G_n)$ is open; since this is the complement of G_n in G it follows that G_n is closed.

For topologies given by sequences of subgroups there is an alternative purely algebraic definition of the completion which is often convenient. Suppose (x_ν) is a Cauchy sequence in G. Then the image of x_ν in G/G_n is ultimately constant, equal say to ξ_n. If we pass from $n + 1$ to n it is clear that $\xi_{n+1} \mapsto \xi_n$ under the projection

$$G/G_{n+1} \xrightarrow{\theta_{n+1}} G/G_n.$$

Thus a Cauchy sequence (x_ν) in G defines a *coherent sequence* (ξ_n) in the sense that

$$\theta_{n+1}\xi_{n+1} = \xi_n \quad \text{for all } n.$$

Moreover it is clear that equivalent Cauchy sequences define the same sequence (ξ_n). Finally, given any coherent sequence (ξ_n), we can construct a Cauchy sequence (x_n) giving rise to it by taking x_n to be any element in the coset ξ_n (so that $x_{n+1} - x_n \in G_n$). Thus $\hat{G}$ can equally well be defined as the set of coherent sequences (ξ_n) with the obvious group structure.

We have now arrived at a special case of *inverse limits*. More generally, consider any sequence of groups $\{A_n\}$ and homomorphisms

$$\theta_{n+1}: A_{n+1} \to A_n.$$

We call this an *inverse system*, and the group of all coherent sequences (a_n) (i.e., $a_n \in A_n$ and $\theta_{n+1}a_{n+1} = a_n$) is called the *inverse limit* of the system. It is usually written $\varprojlim A_n$, the homomorphisms θ_n being understood. With this notation we have

$$\hat{G} \cong \varprojlim G/G_n.$$

The inverse limit definition of $\hat{G}$ has many advantages. Its main drawback is that it presupposes a fixed choice of the subgroups G_n. Now we can have different sequences of G_n defining the same topology and hence the same completion. Of course we could define notions of "equivalent" inverse systems but the merit of the topological language is precisely that such notions are already built into it.

The exactness properties of completions are best studied by inverse limits. First let us observe that the inverse system $\{G/G_n\}$ has the special property that θ_{n+1} is always surjective. Any inverse system with this property we shall call a *surjective* system. Suppose now that $\{A_n\}$, $\{B_n\}$, $\{C_n\}$ are three inverse systems and that we have commutative diagrams of exact sequences

$$\begin{array}{ccccccccc} 0 & \to & A_{n+1} & \to & B_{n+1} & \to & C_{n+1} & \to & 0 \\ & & \downarrow & & \downarrow & & \downarrow & & \\ 0 & \to & A_n & \to & B_n & \to & C_n & \to & 0. \end{array}$$

We shall then say that we have an exact sequence of inverse systems. The diagram certainly induces homomorphisms

$$0 \to \varprojlim A_n \to \varprojlim B_n \to \varprojlim C_n \to 0$$

but this sequence is *not* always exact. However, we have

Proposition 10.2. *If* $0 \to \{A_n\} \to \{B_n\} \to \{C_n\} \to 0$ *is an exact sequence of inverse systems then*

$$0 \to \varprojlim A_n \to \varprojlim B_n \to \varprojlim C_n$$

is always exact. If, moreover, $\{A_n\}$ *is a surjective system then*

$$0 \to \varprojlim A_n \to \varprojlim B_n \to \varprojlim C_n \to 0$$

is exact.

Proof. Let $A = \prod_{n=1}^{\infty} A_n$ and define $d^A: A \to A$ by $d^A(a_n) = a_n - \theta_{n+1}(a_{n+1})$. Then $\operatorname{Ker} d^A \cong \varprojlim A_n$. Define B, C and d^B, d^C similarly. The exact sequence of inverse systems then defines a commutative diagram of exact sequences

$$\begin{array}{ccccccccc} 0 & \to & A & \to & B & \to & C & \to & 0 \\ & & \downarrow{\scriptstyle d^A} & & \downarrow{\scriptstyle d^B} & & \downarrow{\scriptstyle d^C} & & \\ 0 & \to & A & \to & B & \to & C & \to & 0 \end{array}$$

and hence by (2.10) an exact sequence

$$0 \to \operatorname{Ker} d^A \to \operatorname{Ker} d^B \to \operatorname{Ker} d^C \to \operatorname{Coker} d^A \to \operatorname{Coker} d^B \to \operatorname{Coker} d^C \to 0.$$

To complete the proof we have only to prove that

$$\{A_n\} \text{ surjective} \Rightarrow d^A \text{ surjective},$$

but this is clear because to show d^A surjective we have only to solve inductively the equations

$$x_n - \theta_{n+1}(x_{n+1}) = a_n$$

for $x_n \in A_n$, given $a_n \in A_n$. ■

Remark. The group $\operatorname{Coker} d^A$ is usually denoted by $\varprojlim^1 A_n$, since it is a derived functor in the sense of homological algebra.

Corollary 10.3. *Let* $0 \to G' \to G \xrightarrow{p} G'' \to 0$ *be an exact sequence of groups. Let* G *have the topology defined by a sequence* $\{G_n\}$ *of subgroups, and give* G', G'' *the induced topologies, i.e. by the sequences* $\{G'_n \cap G_n\}$, $\{pG_n\}$. *Then*

$$0 \to \hat{G}' \to \hat{G} \to \hat{G}'' \to 0$$

is exact.

Proof. Apply (10.2) to the exact sequences

$$0 \to \frac{G'}{G' \cap G_n} \to \frac{G}{G_n} \to \frac{G''}{pG_n} \to 0. \quad \blacksquare$$

In particular we can apply (10.3) with $G' = G_n$, then $G'' = G/G_n$ has the discrete topology so that $\hat{G}'' = G''$. Hence we deduce

Corollary 10.4. *$\hat{G}_n$ is a subgroup of $\hat{G}$ and*

$$\hat{G}/\hat{G}_n \cong G/G_n. \quad \blacksquare$$

Taking inverse limits in (10.4) we deduce

Proposition 10.5. $\hat{\hat{G}} \cong \hat{G}$. ■

If $\phi: G \to \hat{G}$ is an isomorphism we shall say that G is *complete*. Thus (10.5) asserts that the completion of G is complete. Note that our definition of complete includes Hausdorff (by (10.1)).

The most important class of examples of topological groups of the kind we are considering are given by taking $G = A$, $G_n = \mathfrak{a}^n$, where $\mathfrak{a}$ is an ideal in a ring A. The topology so defined on A is called the *$\mathfrak{a}$-adic topology*, or just the *$\mathfrak{a}$-topology*. Since the $\mathfrak{a}^n$ are ideals, it is not hard to check that with this topology A is a *topological ring*, i.e. that the ring operations are continuous. By (10.1) the topology is Hausdorff $\Leftrightarrow \bigcap \mathfrak{a}^n = (0)$. The *completion* $\hat{A}$ of A is again a topological ring; $\phi: A \to \hat{A}$ is a continuous ring homomorphism, whose kernel is $\bigcap \mathfrak{a}^n$.

Likewise for an A-module M: take $G = M$, $G_n = \mathfrak{a}^n M$. This defines the *$\mathfrak{a}$-topology* on M, and the completion $\hat{M}$ of M is a topological $\hat{A}$-module (i.e. $\hat{A} \times \hat{M} \to \hat{M}$ is continuous). If $f: M \to N$ is any A-module homomorphism, then $f(\mathfrak{a}^n M) = \mathfrak{a}^n f(M) \subseteq \mathfrak{a}^n N$, and therefore f is continuous (with respect to the $\mathfrak{a}$-topologies on M and N) and so defines $\hat{f}: \hat{M} \to \hat{N}$.

Examples. 1) $A = k[x]$, where k is a field and x an indeterminate; $\mathfrak{a} = (x)$. Then $\hat{A} = k[[x]]$, the ring of formal power series.

2) $A = \mathbf{Z}$, $\mathfrak{a} = (p)$, p prime. Then $\hat{A}$ is the ring of *p-adic integers*. Its elements are infinite series $\sum_{n=0}^{\infty} a_n p^n$, $0 \leqslant a_n \leqslant p - 1$. We have $p^n \to 0$ as $n \to \infty$.

FILTRATIONS

The $\mathfrak{a}$-topology of an A-module M was defined by taking the submodules $\mathfrak{a}^n M$ as basic neighborhoods of 0, but there are other ways of defining the same topology. An (infinite) chain $M = M_0 \supseteq M_1 \supseteq \cdots \supseteq M_n \supseteq \cdots$, where the M_n are *submodules* of M, is called a *filtration* of M, and denoted by (M_n). It is an *$\mathfrak{a}$-filtration* if $\mathfrak{a} M_n \subseteq M_{n+1}$ for all n, and a *stable $\mathfrak{a}$-filtration* if $\mathfrak{a} M_n = M_{n+1}$ for all sufficiently large n. Thus $(\mathfrak{a}^n M)$ is a stable $\mathfrak{a}$-filtration.

Lemma 10.6. If (M_n), (M'_n) are stable $\mathfrak{a}$-filtrations of M, then they have bounded difference: that is, there exists an integer n_0 such that $M_{n+n_0} \subseteq M'_n$ and $M'_{n+n_0} \subseteq M_n$ for all $n \geqslant 0$. Hence all stable $\mathfrak{a}$-filtrations determine the same topology on M, namely the $\mathfrak{a}$-topology.

Proof. Enough to take $M'_n = \mathfrak{a}^n M$. Since $\mathfrak{a}M_n \subseteq M_{n+1}$ for all n, we have $\mathfrak{a}^n M \subseteq M_n$; also $\mathfrak{a}M_n = M_{n+1}$ for all $n \geqslant n_0$ say, hence $M_{n+n_0} = \mathfrak{a}^n M_{n_0} \subseteq \mathfrak{a}^n M$. ∎

GRADED RINGS AND MODULES

A *graded ring* is a ring A together with a family $(A_n)_{n \geqslant 0}$ of subgroups of the additive group of A, such that $A = \bigoplus_{n=0}^{\infty} A_n$ and $A_m A_n \subseteq A_{m+n}$ for all $m, n \geqslant 0$. Thus A_0 is a subring of A, and each A_n is an A_0-module.

Example. $A = k[x_1, \ldots, x_r]$, A_n = set of all homogeneous polynomials of degree n.

If A is a graded ring, a *graded A-module* is an A-module M together with a family $(M_n)_{n \geqslant 0}$ of subgroups of M such that $M = \bigoplus_{n=0}^{\infty} M_n$ and $A_m M_n \subseteq M_{m+n}$ for all $m, n \geqslant 0$. Thus each M_n is an A_0-module. An element x of M is *homogeneous* if $x \in M_n$ for some n (n = *degree* of x). Any element $y \in M$ can be written uniquely as a finite sum $\sum_n y_n$, where $y_n \in M_n$ for all $n \geqslant 0$, and all but a finite number of the y_n are 0. The non-zero components y_n are called the *homogeneous components* of y.

If M, N are graded A-modules, a *homomorphism of graded A-modules* is an A-module homomorphism $f: M \to N$ such that $f(M_n) \subseteq N_n$ for all $n \geqslant 0$.

If A is a graded ring, let $A_+ = \bigoplus_{n>0} A_n$. A_+ is an ideal of A.

Proposition 10.7. The following are equivalent, for a graded ring A:

i) *A is a Noetherian ring;*

ii) *A_0 is Noetherian and A is finitely generated as an A_0-algebra.*

Proof. i) ⇒ ii). $A_0 \cong A/A_+$, hence is Noetherian. A_+ is an ideal in A, hence is finitely generated, say by $x_1, \ldots, x_s$, which we may take to be homogeneous elements of A, of degrees $k_1, \ldots, k_s$ say (all > 0). Let A' be the subring of A generated by $x_1, \ldots, x_s$ over A_0. We shall show that $A_n \subseteq A'$ for all $n \geqslant 0$, by induction on n. This is certainly true for $n = 0$. Let $n > 0$ and let $y \in A_n$. Since $y \in A_+$, y is a linear combination of the x_i, say $y = \sum_{i=1}^{s} a_i x_i$, where $a_i \in A_{n-k_i}$ (conventionally $A_m = 0$ if $m < 0$). Since each $k_i > 0$, the inductive hypothesis shows that each a_i is a polynomial in the x's with coefficients in A_0. Hence the same is true of y, and therefore $y \in A'$. Hence $A_n \subseteq A'$ and therefore $A = A'$.

ii) ⇒ i): by Hilbert's basis theorem (7.6). ∎

Let A be a ring (not graded), $\mathfrak{a}$ an ideal of A. Then we can form a graded ring $A^* = \bigoplus_{n=0}^{\infty} \mathfrak{a}^n$. Similarly, if M is an A-module and M_n is an $\mathfrak{a}$-filtration of M, then $M^* = \bigoplus_n M_n$ is a graded A^*-module, since $\mathfrak{a}^m M_n \subseteq M_{m+n}$.

If A is Noetherian, $\mathfrak{a}$ is finitely generated, say by $x_1, \ldots, x_r$; then $A^* = A[x_1, \ldots, x_r]$ and is Noetherian by (7.6).

Lemma 10.8. Let A be a Noetherian ring, M a finitely-generated A-module, (M_n) an $\mathfrak{a}$-filtration of M. Then the following are equivalent:

i) *M^* is a finitely-generated A^*-module;*

ii) *The filtration (M_n) is stable.*

Proof. Each M_n is finitely generated, hence so is each $Q_n = \bigoplus_{r=0}^{n} M_r$: this is a *subgroup* of M^* but not (in general) an A^*-*submodule*. However, it generates one, namely

$$M_n^* = M_0 \oplus \cdots \oplus M_n \oplus \mathfrak{a}M_n \oplus \mathfrak{a}^2 M_n \oplus \cdots \oplus \mathfrak{a}^r M_n \oplus \cdots$$

Since Q_n is finitely generated as an A-module, M_n^* is finitely generated as an A^*-module. The M_n^* form an ascending chain, whose union is M^*. Since A^* is Noetherian, M^* is finitely generated as an A^*-module $\Leftrightarrow$ the chain stops, i.e., $M^* = M_{n_0}^*$ for some $n_0 \Leftrightarrow M_{n_0+r} = \mathfrak{a}^r M_{n_0}$ for all $r \geqslant 0 \Leftrightarrow$ the filtration is stable. ∎

Proposition 10.9. (Artin–Rees lemma). *Let A be a Noetherian ring, $\mathfrak{a}$ an ideal in A, M a finitely-generated A-module, (M_n) a stable $\mathfrak{a}$-filtration of M. If M' is a submodule of M, then $(M' \cap M_n)$ is a stable $\mathfrak{a}$-filtration of M'.*

Proof. We have $\mathfrak{a}(M' \cap M_n) \subseteq \mathfrak{a}M' \cap \mathfrak{a}M_n \subseteq M' \cap M_{n+1}$, hence $(M' \cap M_n)$ is an $\mathfrak{a}$-filtration. Hence it defines a graded A^*-module which is a submodule of M^* and therefore finitely generated (since A^* is Noetherian). Now use (10.8). ∎

Taking $M_n = \mathfrak{a}^n M$ we obtain what is usually known as the Artin–Rees lemma:

Corollary 10.10. There exists an integer k such that

$$(\mathfrak{a}^n M) \cap M' = \mathfrak{a}^{n-k}((\mathfrak{a}^k M) \cap M')$$

for all $n \geqslant k$. ∎

On the other hand, combining (10.9) with the elementary lemma (10.6) we obtain the really significant version:

Theorem 10.11. Let A be a Noetherian ring, $\mathfrak{a}$ an ideal, M a finitely-generated A-module and M' a submodule of M. Then the filtrations $\mathfrak{a}^n M'$ and $(\mathfrak{a}^n M) \cap M'$ have bounded difference. In particular the $\mathfrak{a}$-topology of M' coincides with the topology induced by the $\mathfrak{a}$-topology of M. ∎

Remark. In this chapter we shall apply the last part of (10.11) concerning topologies. However, in the next chapter the stronger result about bounded differences will be needed.

As a first application of (10.11) we combine it with (10.3) to get the important *exactness property of completion*:

Proposition 10.12. *Let*

$$0 \to M' \to M \to M'' \to 0$$

be an exact sequence of finitely-generated modules over a Noetherian ring A. Let $\mathfrak{a}$ be an ideal of A, then the sequence of $\mathfrak{a}$-adic completions

$$0 \to \hat{M}' \to \hat{M} \to \hat{M}'' \to 0$$

is exact. ∎

Since we have a natural homomorphism $A \to \hat{A}$ we can regard $\hat{A}$ as an A-algebra and so for any A-module M we can form an $\hat{A}$-module $\hat{A} \otimes_A M$. It is natural to ask how this compares with the $\hat{A}$-module $\hat{M}$. Now the A-module homomorphism $M \to \hat{M}$ defines an $\hat{A}$-module homomorphism

$$\hat{A} \otimes_A M \to \hat{A} \otimes_A \hat{M} \to \hat{A} \otimes_{\hat{A}} \hat{M} = \hat{M}.$$

In general, for arbitrary A and M, this is neither injective nor surjective, but we do have:

Proposition 10.13. *For any ring A, if M is finitely-generated, $\hat{A} \otimes_A M \to \hat{M}$ is surjective. If, moreover, A is Noetherian then $\hat{A} \otimes_A M \to \hat{M}$ is an isomorphism.*

Proof. Using (10.3) or otherwise it is clear that $\mathfrak{a}$-adic completion commutes with finite direct sums. Hence if $F \cong A^n$ we have $\hat{A} \otimes_A F \cong \hat{F}$. Now assume M is finitely generated so that we have an exact sequence

$$0 \to N \to F \to M \to 0.$$

This gives rise to the commutative diagram

$$\begin{array}{ccccccc} & \hat{A} \otimes_A N & \to & \hat{A} \otimes_A F & \to & \hat{A} \otimes_A M & \to 0 \\ & \downarrow{\scriptstyle\gamma} & & \downarrow{\scriptstyle\beta} & & \downarrow{\scriptstyle\alpha} & \\ 0 \to & \hat{N} & \to & \hat{F} & \xrightarrow{\delta} & \hat{M} & \to 0 \end{array}$$

in which the top line is exact (by (2.18)). By (10.3) δ is surjective. Since β is an isomorphism this implies that α is surjective, proving the first part of the proposition. Assume now that A is Noetherian, then N is also finitely generated so that γ is surjective and, by (10.12), the bottom line is exact. A little diagram chasing now proves that α is injective and so an isomorphism. ∎

Propositions (10.12) and (10.13) together assert that the functor $M \mapsto \hat{A} \otimes_A M$ is exact on the category of finitely-generated A-modules (when A is Noetherian). As shown in Chapter 2 this proves:

Proposition 10.14. *If A is a Noetherian ring, $\mathfrak{a}$ an ideal, $\hat{A}$ the $\mathfrak{a}$-adic completion of A, then $\hat{A}$ is a flat A-algebra.* ■

Remark. For non-finitely-generated modules the functor $M \mapsto \hat{M}$ is not exact: the good functor, which is exact, is $M \mapsto \hat{A} \otimes_A M$ and the two functors coincide on finitely-generated modules.

We proceed now to study the ring $\hat{A}$ in more detail. First some elementary propositions:

Proposition 10.15. *If A is Noetherian, $\hat{A}$ its $\mathfrak{a}$-adic completion, then*

i) $\hat{\mathfrak{a}} = \hat{A}\mathfrak{a} \cong \hat{A} \otimes_A \mathfrak{a}$;

ii) $(\mathfrak{a}^n)^\wedge = (\hat{\mathfrak{a}})^n$;

iii) $\mathfrak{a}^n/\mathfrak{a}^{n+1} \cong \hat{\mathfrak{a}}^n/\hat{\mathfrak{a}}^{n+1}$;

iv) *$\hat{\mathfrak{a}}$ is contained in the Jacobson radical of $\hat{A}$.*

Proof. Since A is Noetherian, $\mathfrak{a}$ is finitely-generated. (10.13) implies that the map

$$\hat{A} \otimes_A \mathfrak{a} \to \hat{\mathfrak{a}},$$

whose image is $\hat{A}\mathfrak{a}$, is an isomorphism. This proves i). Now apply i) to $\mathfrak{a}^n$ and we deduce that

$$\begin{aligned}(\mathfrak{a}^n)^\wedge &= \hat{A}\mathfrak{a}^n = (\hat{A}\mathfrak{a})^n && \text{by (1.18)}\\ &= (\hat{\mathfrak{a}})^n && \text{by i).}\end{aligned}$$

Applying (10.4) we now deduce

$$A/\mathfrak{a}^n \cong \hat{A}/\hat{\mathfrak{a}}^n$$

from which iii) follows by taking quotients. By ii) and (10.5) we see that $\hat{A}$ is complete for its $\hat{\mathfrak{a}}$-topology. Hence for any $x \in \hat{\mathfrak{a}}$

$$(1 - x)^{-1} = 1 + x + x^2 + \cdots$$

converges in $\hat{A}$, so that $1 - x$ is a unit. By (1.9) this implies that $\hat{\mathfrak{a}}$ is contained in the Jacobson radical of $\hat{A}$. ■

Proposition 10.16. *Let A be a Noetherian local ring, $\mathfrak{m}$ its maximal ideal. Then the $\mathfrak{m}$-adic completion $\hat{A}$ of A is a local ring with maximal ideal $\hat{\mathfrak{m}}$.*

Proof. By (10.15) iii) we have $\hat{A}/\hat{\mathfrak{m}} \cong A/\mathfrak{m}$, hence $\hat{A}/\hat{\mathfrak{m}}$ is a field and so $\hat{\mathfrak{m}}$ is a maximal ideal. By (10.15) iv) it follows that $\hat{\mathfrak{m}}$ is the Jacobson radical of $\hat{A}$ and so is the unique maximal ideal. Thus $\hat{A}$ is a local ring. ■

The important question of how much we lose on completion is answered by Krull's Theorem:

Theorem 10.17. *Let A be a Noetherian ring, $\mathfrak{a}$ an ideal, M a finitely-generated A-module and $\hat{M}$ the $\mathfrak{a}$-completion of M. Then the kernel $E = \bigcap_{n=1}^{\infty} \mathfrak{a}^n M$ of $M \to \hat{M}$ consists of those $x \in M$ annihilated by some element of $1 + \mathfrak{a}$.*

Proof. Since E is the intersection of all neighborhoods of $0 \in M$, the topology induced on it is trivial, i.e., E is the only neighborhood of $0 \in E$. By (10.11) the induced topology on E coincides with its $\mathfrak{a}$-topology. Since $\mathfrak{a}E$ is a neighborhood in the $\mathfrak{a}$-topology it follows that $\mathfrak{a}E = E$. Since M is finitely-generated and A is Noetherian, E is also finitely-generated and so we can apply (2.5) and deduce from $\mathfrak{a}E = E$ that $(1 - \alpha)E = 0$ for some $\alpha \in \mathfrak{a}$. The converse is obvious: if $(1 - \alpha)x = 0$, then

$$x = \alpha x = \alpha^2 x = \cdots \in \bigcap_{n=1}^{\infty} \mathfrak{a}^n M = E. \quad \blacksquare$$

Remarks. 1) If S is the multiplicatively closed set $1 + \mathfrak{a}$, then (10.17) asserts that

$$A \to \hat{A} \text{ and } A \to S^{-1}A$$

have the same kernel. Moreover for any $\alpha \in \hat{\mathfrak{a}}$

$$(1 - \alpha)^{-1} = 1 + \alpha + \alpha^2 + \cdots$$

converges in $\hat{A}$, so that every element of S becomes a unit in $\hat{A}$. By the universal property of $S^{-1}A$ this means that there is a natural homomorphism $S^{-1}A \to \hat{A}$ and (10.17) implies that this is injective. Thus $S^{-1}A$ can be identified with a subring of $\hat{A}$.

2) Krull's Theorem (10.17) may be false if A is not Noetherian. Let A be the ring of all C^{∞} functions on the real line, and let $\mathfrak{a}$ be the ideal of all f which vanish at the origin ($\mathfrak{a}$ is maximal since $A/\mathfrak{a} \cong \mathbf{R}$). In fact $\mathfrak{a}$ is generated by the identity function x, and $\bigcap_{n=1}^{\infty} \mathfrak{a}^n$ is the set of all $f \in A$, all of whose derivatives vanish at the origin. On the other hand f is annihilated by some element $1 + \alpha$ ($\alpha \in \mathfrak{a}$) if and only if f vanishes identically in some neighborhood of 0. The well-known function e^{-1/x^2}, which is not identically zero near 0, but has vanishing derivatives at 0, then shows that the kernels of

$$A \to \hat{A} \text{ and } A \to S^{-1}A \qquad (S = 1 + \mathfrak{a})$$

do not coincide. Thus A is not Noetherian.

Krull's Theorem has many corollaries:

Corollary 10.18. *Let A be a Noetherian domain, $\mathfrak{a} \neq (1)$ an ideal of A. Then $\bigcap \mathfrak{a}^n = 0$.*

Proof. $1 + \mathfrak{a}$ contains no zero-divisors. $\blacksquare$

Corollary 10.19. *Let A be a Noetherian ring, $\mathfrak{a}$ an ideal of A contained in the Jacobson radical and let M be a finitely-generated A-module. Then the $\mathfrak{a}$-topology of M is Hausdorff,* i.e. $\bigcap \mathfrak{a}^n M = 0$.

Proof. By (1.9) every element of $1 + \mathfrak{a}$ is a unit. ∎

As a particularly important special case of (10.19) we have:

Corollary 10.20. *Let A be a Noetherian local ring, $\mathfrak{m}$ its maximal ideal, M a finitely-generated A-module. Then the $\mathfrak{m}$-topology of M is Hausdorff. In particular the $\mathfrak{m}$-topology of A is Hausdorff.* ∎

We can restate (10.20) slightly differently if we recall that an $\mathfrak{m}$-primary ideal of A is just any ideal contained between $\mathfrak{m}$ and some power $\mathfrak{m}^n$ (use (4.2) and (7.14)). Thus (10.20) implies that the intersection of all $\mathfrak{m}$-primary ideals of A is zero. If now A is any Noetherian ring, $\mathfrak{p}$ a prime ideal, we can apply this version of (10.20) to the local ring $A_{\mathfrak{p}}$. Lifting back to A and using the one-to-one correspondence (4.8) between $\mathfrak{p}$-primary ideals of A and $\mathfrak{m}$-primary ideals of $A_{\mathfrak{p}}$ (where $\mathfrak{m} = \mathfrak{p}A_{\mathfrak{p}}$) we deduce:

Corollary 10.21. *Let A be a Noetherian ring, $\mathfrak{p}$ a prime ideal of A. Then the intersection of all $\mathfrak{p}$-primary ideals of A is the kernel of $A \to A_{\mathfrak{p}}$.* ∎

THE ASSOCIATED GRADED RING

Let A be a ring and $\mathfrak{a}$ an ideal of A. Define

$$G(A)\left(= G_{\mathfrak{a}}(A)\right) = \bigoplus_{n=0}^{\infty} \mathfrak{a}^n/\mathfrak{a}^{n+1} \qquad (\mathfrak{a}^0 = A).$$

This is a graded ring, in which the multiplication is defined as follows: For each $x_n \in \mathfrak{a}^n$, let $\bar{x}_n$ denote the image of x_n in $\mathfrak{a}^n/\mathfrak{a}^{n+1}$; define $\bar{x}_m\bar{x}_n$ to be $\overline{x_m x_n}$, i.e., the image of $x_m x_n$ in $\mathfrak{a}^{m+n}/\mathfrak{a}^{m+n+1}$; check that $\bar{x}_m\bar{x}_n$ does not depend on the particular representatives chosen.

Similarly, if M is an A-module and (M_n) is an $\mathfrak{a}$-filtration of M, define

$$G(M) = \bigoplus_{n=0}^{\infty} M_n/M_{n+1}$$

which is a graded $G(A)$-module in a natural way. Let $G_n(M)$ denote M_n/M_{n+1}.

Proposition 10.22. *Let A be a Noetherian ring, $\mathfrak{a}$ an ideal of A. Then*

i) *$G_{\mathfrak{a}}(A)$ is Noetherian;*

ii) *$G_{\mathfrak{a}}(A)$ and $G_{\hat{\mathfrak{a}}}(\hat{A})$ are isomorphic as graded rings;*

iii) *if M is a finitely-generated A-module and (M_n) is a stable $\mathfrak{a}$-filtration of M, then $G(M)$ is a finitely-generated graded $G_{\mathfrak{a}}(A)$-module.*

Proof. i) Since A is Noetherian, $\mathfrak{a}$ is finitely generated, say by $x_1, \ldots, x_s$. Let $\bar{x}_i$ be the image of x_i in $\mathfrak{a}/\mathfrak{a}^2$, then $G(A) = (A/\mathfrak{a})[\bar{x}_1, \ldots, \bar{x}_s]$. Since $A/\mathfrak{a}$ is Noetherian, $G(A)$ is Noetherian by the Hilbert basis theorem.

ii) $\mathfrak{a}^n/\mathfrak{a}^{n+1} \cong \hat{\mathfrak{a}}^n/\hat{\mathfrak{a}}^{n+1}$ by (10.15) iii).

iii) There exists n_0 such that $M_{n_0+r} = \mathfrak{a}^r M_{n_0}$ for all $r \geqslant 0$, hence $G(M)$ is generated by $\bigoplus_{n \leqslant n_0} G_n(M)$. Each $G_n(M) = M_n/M_{n+1}$ is Noetherian and annihilated by $\mathfrak{a}$, hence is a finitely-generated $A/\mathfrak{a}$-module, hence $\bigoplus_{n \leqslant n_0} G_n(M)$ is generated by a finite number of elements (as an $A/\mathfrak{a}$-module), hence $G(M)$ is finitely generated as a $G(A)$-module. ∎

The last main result of this chapter is that the $\mathfrak{a}$-adic completion of a Noetherian ring is Noetherian. Before we can proceed to the proof we need a simple lemma connecting the completion of any filtered group and the associated graded group.

Lemma 10.23. *Let $\phi: A \to B$ be a homomorphism of filtered groups,* i.e. *$\phi(A_n) \subseteq B_n$, and let $G(\phi): G(A) \to G(B)$, $\hat{\phi}: \hat{A} \to \hat{B}$ be the induced homomorphisms of the associated graded and completed groups. Then*

i) *$G(\phi)$ injective $\Rightarrow$ $\hat{\phi}$ injective;*

ii) *$G(\phi)$ surjective $\Rightarrow$ $\hat{\phi}$ surjective.*

Proof. Consider the commutative diagram of exact sequences

$$\begin{array}{ccccccccc} 0 & \to & A_n/A_{n+1} & \to & A/A_{n+1} & \to & A/A_n & \to & 0 \\ & & \downarrow{\scriptstyle G_n(\phi)} & & \downarrow{\scriptstyle \alpha_{n+1}} & & \downarrow{\scriptstyle \alpha_n} & & \\ 0 & \to & B_n/B_{n+1} & \to & B/B_{n+1} & \to & B/B_n & \to & 0. \end{array}$$

This gives the exact sequence

$$0 \to \operatorname{Ker} G_n(\phi) \to \operatorname{Ker} \alpha_{n+1} \to \operatorname{Ker} \alpha_n \to \operatorname{Coker} G_n(\phi) \to \operatorname{Coker} \alpha_{n+1} \to \operatorname{Coker} \alpha_n \to 0.$$

From this we see, by induction on n, that $\operatorname{Ker} \alpha_n = 0$ (case i)) or $\operatorname{Coker} \alpha_n = 0$ (case ii)). Moreover in case ii) we also have $\operatorname{Ker} \alpha_{n+1} \to \operatorname{Ker} \alpha_n$ surjective. Taking the inverse limit of the homomorphisms α_n and applying (10.2) the lemma follows. ∎

We can now form a result which is a partial converse of (10.22) iii) and is the main step in showing that $\hat{A}$ is Noetherian.

Proposition 10.24. *Let A be a ring, $\mathfrak{a}$ an ideal of A, M an A-module, (M_n) an $\mathfrak{a}$-filtration of M. Suppose that A is complete in the $\mathfrak{a}$-topology and that M is Hausdorff in its filtration topology* (i.e. *that $\bigcap_n M_n = 0$). Suppose also that $G(M)$ is a finitely-generated $G(A)$-module. Then M is a finitely-generated A-module.*

Proof. Pick a finite set of generators of $G(M)$, and split them up into their homogeneous components, say ξ_i $(1 \leqslant i \leqslant \nu)$ where ξ_i has degree say $n(i)$, and is therefore the image of say $x_i \in M_{n(i)}$. Let F^i be the module A with the stable $\mathfrak{a}$-filtration given by $F^i_k = \mathfrak{a}^{k+n(i)}$ and put $F = \bigoplus_{i=1}^{\nu} F^i$. Then mapping the generator 1 of each F^i to x_i defines a homomorphism

$$\phi: F \to M$$

of filtered groups, and $G(\phi): G(F) \to G(M)$ is a homomorphism of $G(A)$ modules. By construction it is surjective. Hence by (10.23) ii) $\hat{\phi}$ is surjective. Consider now the diagram

$$\begin{array}{ccc} F & \xrightarrow{\phi} & M \\ {\scriptstyle\alpha}\downarrow & & \downarrow{\scriptstyle\beta} \\ \hat{F} & \xrightarrow{\hat{\phi}} & \hat{M} \end{array}$$

Since F is free and $A = \hat{A}$ it follows that α is an isomorphism. Since M is Hausdorff β is injective. The surjectivity of $\hat{\phi}$ thus implies the surjectivity of ϕ, and this means that $x_1, \ldots, x_r$ generate M as an A-module. ∎

Corollary 10.25. *With the hypotheses of* (10.24), *if* $G(M)$ *is a Noetherian* $G(A)$*-module, then* M *is a Noetherian* A*-module.*

Proof. We have to show that every submodule M' of M is finitely generated (6.2). Let $M'_n = M' \cap M_n$; then (M'_n) is an $\mathfrak{a}$-filtration of M', and the embedding $M'_n \to M_n$ gives rise to an injective homomorphism $M'_n/M'_{n+1} \to M_n/M_{n+1}$, hence to an embedding of $G(M')$ in $G(M)$. Since $G(M)$ is Noetherian, $G(M')$ is finitely generated by (6.2); also M' is Hausdorff, since $\bigcap M'_n \subseteq \bigcap M_n = 0$; hence by (10.24) M' is finitely generated. ∎

We can now deduce the result we are after:

Theorem 10.26. *If* A *is a Noetherian ring,* $\mathfrak{a}$ *an ideal of* A*, then the* $\mathfrak{a}$*-completion* $\hat{A}$ *of* A *is Noetherian.*

Proof. By (10.22) we know that

$$G_{\mathfrak{a}}(A) = G_{\hat{\mathfrak{a}}}(\hat{A})$$

is Noetherian. Now apply (10.25) to the complete ring $\hat{A}$, taking $M = \hat{A}$ (filtered by $\hat{\mathfrak{a}}^n$, and so Hausdorff). ∎

Corollary 10.27. *If* A *is a Noetherian ring, the power series ring* $B = A[[x_1, \ldots, x_n]]$ *in* n *variables is Noetherian. In particular* $k[[x_1, \ldots, x_n]]$ *(*k *a field) is Noetherian.*

Proof. $A[x_1, \ldots, x_n]$ is Noetherian by the Hilbert basis theorem, and B is its completion for the $(x_1, \ldots, x_n)$-adic topology. ∎

EXERCISES

1. Let $\alpha_n: \mathbf{Z}/p\mathbf{Z} \to \mathbf{Z}/p^n\mathbf{Z}$ be the injection of abelian groups given by $\alpha_n(1) = p^{n-1}$, and let $\alpha: A \to B$ be the direct sum of all the α_n (where A is a countable direct sum of copies of $\mathbf{Z}/p\mathbf{Z}$, and B is the direct sum of the $\mathbf{Z}/p^n\mathbf{Z}$). Show that the p-adic completion of A is just A but that the completion of A for the topology induced from the p-adic topology on B is the direct *product* of the $\mathbf{Z}/p\mathbf{Z}$. Deduce that p-adic completion is *not* a right-exact functor on the category of all $\mathbf{Z}$-modules.

2. In Exercise 1, let $A_n = \alpha^{-1}(p^nB)$, and consider the exact sequence

$$0 \to A_n \to A \to A/A_n \to 0.$$

Show that $\varprojlim$ is not right exact, and compute $\varprojlim^1 A_n$.

3. Let A be a Noetherian ring, $\mathfrak{a}$ an ideal and M a finitely-generated A-module. Using Krull's Theorem and Exercise 14 of Chapter 3, prove that

$$\bigcap_{n=1}^{\infty} \mathfrak{a}^n M = \bigcap_{\mathfrak{m} \supseteq \mathfrak{a}} \operatorname{Ker}(M \to M_{\mathfrak{m}}),$$

where $\mathfrak{m}$ runs over all maximal ideals containing $\mathfrak{a}$.

Deduce that

$$\hat{M} = 0 \Leftrightarrow \operatorname{Supp}(M) \cap V(\mathfrak{a}) = \emptyset \qquad (\text{in } \operatorname{Spec}(A)).$$

[The reader should think of $\hat{M}$ as the "Taylor expansion" of M transversal to the subscheme $V(\mathfrak{a})$: the above result then shows that M is determined in a neighborhood of $V(\mathfrak{a})$ by its Taylor expansion.]

4. Let A be a Noetherian ring, $\mathfrak{a}$ an ideal in A, and $\hat{A}$ the $\mathfrak{a}$-adic completion. For any $x \in A$, let $\hat{x}$ be the image of x in $\hat{A}$. Show that

$$x \text{ not a zero-divisor in } A \Rightarrow \hat{x} \text{ not a zero-divisor in } \hat{A}.$$

Does this imply that

$$A \text{ is an integral domain} \Rightarrow \hat{A} \text{ is an integral domain?}$$

[Apply the exactness of completion to the sequence $0 \to A \xrightarrow{x} A$.]

5. Let A be a Noetherian ring and let $\mathfrak{a}, \mathfrak{b}$ be ideals in A. If M is any A-module, let $M^{\mathfrak{a}}, M^{\mathfrak{b}}$ denote its $\mathfrak{a}$-adic and $\mathfrak{b}$-adic completions respectively. If M is finitely generated, prove that $(M^{\mathfrak{a}})^{\mathfrak{b}} \cong M^{\mathfrak{a}+\mathfrak{b}}$.
[Take the $\mathfrak{a}$-adic completion of the exact sequence

$$0 \to \mathfrak{b}^m M \to M \to M/\mathfrak{b}^m M \to 0$$

and apply (10.13). Then use the isomorphism

$$\varprojlim_m (\varprojlim_n M/(\mathfrak{a}^n M + \mathfrak{b}^m M)) \cong \varprojlim_n M/(\mathfrak{a}^n M + \mathfrak{b}^n M)$$

and the inclusions $(\mathfrak{a} + \mathfrak{b})^{2n} \subseteq \mathfrak{a}^n + \mathfrak{b}^n \subseteq (\mathfrak{a} + \mathfrak{b})^n$.]

6. Let A be a Noetherian ring and $\mathfrak{a}$ an ideal in A. Prove that $\mathfrak{a}$ is contained in the Jacobson radical of A if and only if every maximal ideal of A is closed for the $\mathfrak{a}$-topology. (A Noetherian topological ring in which the topology is defined by an ideal contained in the Jacobson radical is called a *Zariski ring*. Examples are local rings and (by (10.15)(iv)) $\mathfrak{a}$-adic completions.)

7. Let A be a Noetherian ring, $\mathfrak{a}$ an ideal of A, and $\hat{A}$ the $\mathfrak{a}$-adic completion. Prove that $\hat{A}$ is faithfully flat over A (Chapter 3, Exercise 16) if and only if A is a Zariski ring (for the $\mathfrak{a}$-topology).

[Since $\hat{A}$ is flat over A, it is enough to show that

$$M \to \hat{M} \text{ injective for all finitely generated } M \Leftrightarrow A \text{ is Zariski;}$$

now use (10.19) and Exercise 6.]

8. Let A be the local ring of the origin in $\mathbf{C}^n$ (i.e., the ring of all rational functions $f/g \in \mathbf{C}(z_1, \ldots, z_n)$ with $g(0) \neq 0$), let B be the ring of power series in $z_1, \ldots, z_n$ which converge in some neighborhood of the origin, and let C be the ring of formal power series in $z_1, \ldots, z_n$, so that $A \subset B \subset C$. Show that B is a local ring and that its completion for the maximal ideal topology is C. Assuming that B is Noetherian, prove that B is A-flat. [Use Chapter 3, Exercise 17, and Exercise 7 above.]

9. Let A be a local ring, $\mathfrak{m}$ its maximal ideal. Assume that A is $\mathfrak{m}$-adically complete. For any polynomial $f(x) \in A[x]$, let $\bar{f}(x) \in (A/\mathfrak{m})[x]$ denote its reduction mod. $\mathfrak{m}$. Prove *Hensel's lemma*: if $f(x)$ is monic of degree n and if there exist coprime monic polynomials $\bar{g}(x)$, $\bar{h}(x) \in (A/\mathfrak{m})[x]$ of degrees $r, n - r$ with $\bar{f}(x) = \bar{g}(x)\bar{h}(x)$, then we can lift $\bar{g}(x)$, $\bar{h}(x)$ back to monic polynomials $g(x)$, $h(x) \in A[x]$ such that $f(x) = g(x)h(x)$.
[Assume inductively that we have constructed $g_k(x)$, $h_k(x) \in A[x]$ such that $g_k(x)h_k(x) - f(x) \in \mathfrak{m}^k A[x]$. Then use the fact that since $\bar{g}(x)$ and $\bar{h}(x)$ are coprime we can find $\bar{a}_p(x)$, $\bar{b}_p(x)$, of degrees $\leqslant n - r, r$ respectively, such that $x^p = \bar{a}_p(x)\bar{g}_k(x) + \bar{b}_p(x)\bar{h}_k(x)$, where p is any integer such that $1 \leqslant p \leqslant n$. Finally, use the completeness of A to show that the sequences $g_k(x)$, $h_k(x)$ converge to the required $g(x)$, $h(x)$.]

10. i) With the notation of Exercise 9, deduce from Hensel's lemma that if $\bar{f}(x)$ has a simple root $\alpha \in A/\mathfrak{m}$, then $f(x)$ has a simple root $a \in A$ such that $\alpha = a \bmod \mathfrak{m}$.
ii) Show that 2 is a square in the ring of 7-adic integers.
iii) Let $f(x, y) \in k[x, y]$, where k is a field, and assume that $f(0, y)$ has $y = a_0$ as a simple root. Prove that there exists a formal power series $y(x) = \sum_{n=0}^{\infty} a_n x^n$ such that $f(x, y(x)) = 0$.
(This gives the "analytic branch" of the curve $f = 0$ through the point $(0, a_0)$.)

11. Show that the converse of (10.26) is false, even if we assume that A is local and that $\hat{A}$ is a finitely-generated A-module.
[Take A to be the ring of germs of C^∞ functions of x at $x = 0$, and use Borel's Theorem that every power series occurs as the Taylor expansion of some C^∞ function.]

12. If A is Noetherian, then $A[[x_1, \ldots, x_n]]$ is a faithfully flat A-algebra. [Express $A \to A[[x_1, \ldots, x_n]]$ as a composition of flat extensions, and use Exercise 5(v) of Chapter 1.]

11

Dimension Theory

One of the basic notions in algebraic geometry is that of the dimension of a variety. This is essentially a local notion, and, as we shall show in this chapter, there is a very satisfactory theory of dimension for general Noetherian local rings. The main theorem asserts the equivalence of three different definitions of dimension. Two of these definitions have a fairly obvious geometrical content, but the third involving the Hilbert function is less conceptual. It has, however, many technical advantages and the whole theory becomes more streamlined if one brings it in at an early stage.

After dealing with dimension we give a brief account of regular local rings, which correspond to the notion of non-singularity in algebraic geometry. We establish the equivalence of three definitions of regularity.

Finally we indicate how, in the case of algebraic varieties over a field, the local dimensions we have defined coincide with the transcendence degree of the function field.

HILBERT FUNCTIONS

Let $A = \bigoplus_{n=0}^{\infty} A_n$ be a Noetherian graded ring. By (10.7) A_0 is a Noetherian ring, and A is generated (as an A_0-algebra) by say $x_1, \ldots, x_s$, which we may take to be homogeneous, of degrees $k_1, \ldots, k_s$ (all > 0).

Let M be a finitely-generated graded A-module. Then M is generated by a finite number of homogeneous elements, say m_j $(1 \leqslant j \leqslant t)$; let $r_j = \deg m_j$. Every element of M_n, the homogeneous component of M of degree n, is thus of the form $\sum_j f_j(x) m_j$, where $f_j(x) \in A$ is homogeneous of degree $n - r_j$ (and therefore zero if $n < r_j$). It follows that M_n is finitely generated as an A_0-module, namely it is generated by all $g_j(x) m_j$ where $g_j(x)$ is a monomial in the x_i of total degree $n - r_j$.

Let λ be an *additive function* (with values in $\mathbf{Z}$) on the class of all finitely-generated A_0-modules (Chapter 2). The *Poincaré series* of M (with respect to λ) is the generating function of $\lambda(M_n)$, i.e., it is the power series

$$P(M, t) = \sum_{n=0}^{\infty} \lambda(M_n) t^n \quad \in \mathbf{Z}[[t]].$$

Theorem 11.1. (Hilbert, Serre). *$P(M, t)$ is a rational function in t of the form $f(t)/\prod_{i=1}^{s}(1 - t^{k_i})$, where $f(t) \in \mathbf{Z}[t]$.*

Proof. By induction on s, the number of generators of A over A_0. Start with $s = 0$; this means that $A_n = 0$ for all $n > 0$, so that $A = A_0$ and M is a finitely-generated A_0 module, hence $M_n = 0$ for all large n. Thus $P(M, t)$ is a polynomial in this case.

Now suppose $s > 0$ and the theorem true for $s - 1$. Multiplication by x_s is an A-module homomorphism of M_n into M_{n+k_s}, hence it gives an exact sequence, say

$$0 \to K_n \to M_n \xrightarrow{x_s} M_{n+k_s} \to L_{n+k_s} \to 0. \tag{1}$$

Let $K = \bigoplus_n K_n$, $L = \bigoplus_n L_n$; these are both finitely-generated A-modules (because K is a submodule and L a quotient module of M), and both are annihilated by x_s, hence they are $A_0[x_1, \ldots, x_{s-1}]$-modules. Applying λ to (1) we have, by (2.11)

$$\lambda(K_n) - \lambda(M_n) + \lambda(M_{n+k_s}) - \lambda(L_{n+k_s}) = 0;$$

multiplying by t^{n+k_s} and summing with respect to n we get

$$(1 - t^{k_s})P(M, t) = P(L, t) - t^{k_s}P(K, t) + g(t) \tag{2}$$

where $g(t)$ is a polynomial. Applying the inductive hypothesis the result now follows. ■

The order of the pole of $P(M, t)$ at $t = 1$ we shall denote by $d(M)$. It provides a measure of the "size" of M (relative to λ). In particular $d(A)$ is defined. The case when all $k_i = 1$ is specially simple:

Corollary 11.2. If each $k_i = 1$, then for all sufficiently large n, $\lambda(M_n)$ is a polynomial in n (with rational coefficients) of degree $d - 1$.*

Proof. By (11.1) we have $\lambda(M_n) =$ coefficient of t^n in $f(t)\cdot(1 - t)^{-s}$. Canceling powers of $(1 - t)$ we may assume $s = d$ and $f(1) \neq 0$. Suppose $f(t) = \sum_{k=0}^{N} a_k t^k$; since

$$(1 - t)^{-d} = \sum_{k=0}^{\infty} \binom{d+k-1}{d-1} t^k$$

we have

$$\lambda(M_n) = \sum_{k=0}^{N} a_k \binom{d+n-k-1}{d-1} \text{ for all } n \geqslant N.$$

and the sum on the right-hand side is a polynomial in n with leading term $(\sum a_k)n^{d-1}/(d - 1)! \neq 0$. ■

Remarks. 1) For a polynomial $f(x)$ to be such that $f(n)$ is an integer for all integers n, it is not necessary for f to have integer coefficients: e.g., $\frac{1}{2}x(x + 1)$.

* We adopt the convention here that the degree of the zero polynomial is -1: also that the binomial coefficient $\binom{n}{-1} = 0$ for $n \geq 0$, and $= 1$ for $n = -1$.

2) The polynomial in (11.2) is usually called the *Hilbert function* (or polynomial) of M (with respect to λ).

Returning now to the sequence (1) let us replace x_s by any element $x \in A_k$ which is not a zero-divisor in M (i.e., $xm = 0$ with $m \in M \Rightarrow m = 0$). Then $K = 0$ and equation (2) shows that

$$d(L) = d(M) - 1.$$

Thus we have proved

Proposition 11.3. *If $x \in A_k$ is not a zero-divisor in M then $d(M/xM) = d(M) - 1$.* ■

We shall use (11.1) in the case where A_0 is an *Artin* ring (in particular, a field) and $\lambda(M)$ is the *length* $l(M)$ of a finitely-generated A_0-module M. By (6.9) $l(M)$ is additive.

Example. Let $A = A_0[x_1, \ldots, x_s]$, where A_0 is an Artin ring and the x_i are independent indeterminates. Then A_n is a free A_0-module generated by the monomials $x_1^{m_1} \cdots x_s^{m_s}$ where $\sum m_i = n$; there are $\binom{s+n-1}{s-1}$ of these, hence $P(A, t) = (1 - t)^{-s}$.

We shall now consider the Hilbert functions obtained from a local ring by passing to the associated graded rings as in Chapter 10.

Proposition 11.4. *Let A be a Noetherian local ring, $\mathfrak{m}$ its maximal ideal, $\mathfrak{q}$ an $\mathfrak{m}$-primary ideal, M a finitely-generated A-module, (M_n) a stable $\mathfrak{q}$-filtration of M. Then*

i) M/M_n is of finite length, for each $n \geqslant 0$;

ii) for all sufficiently large n this length is a polynomial $g(n)$ of degree $\leqslant s$ in n, where s is the least number of generators of $\mathfrak{q}$;

iii) the degree and leading coefficient of $g(n)$ depend only on M and $\mathfrak{q}$, not on the filtration chosen.

Proof. i) Let $G(A) = \bigoplus_n \mathfrak{q}^n/\mathfrak{q}^{n+1}$, $G(M) = \bigoplus_n M_n/M_{n+1}$. $G_0(A) = A/\mathfrak{q}$ is an Artin local ring, say by (8.5); $G(A)$ is Noetherian, and $G(M)$ is a finitely-generated graded $G(A)$-module (10.22). Each $G_n(M) = M_n/M_{n+1}$ is a Noetherian A-module annihilated by $\mathfrak{q}$, hence a Noetherian $A/\mathfrak{q}$-module, and therefore of finite length (since $A/\mathfrak{q}$ is Artin). Hence M/M_n is of finite length, and

$$l_n = l(M/M_n) = \sum_{r=1}^{n} l(M_{r-1}/M_r). \tag{1}$$

ii) If $x_1, \ldots, x_s$ generate $\mathfrak{q}$, the images $\bar{x}_i$ of the x_i in $\mathfrak{q}/\mathfrak{q}^2$ generate $G(A)$ as an $A/\mathfrak{q}$-algebra, and each $\bar{x}_i$ has degree 1. Hence by (11.2) we have $l(M_n/M_{n+1}) = f(n)$ say, where $f(n)$ is a polynomial in n of degree $\leqslant s - 1$ for all large n.

Since from (1) we have $l_{n+1} - l_n = f(n)$, it follows that l_n is a polynomial $g(n)$ of degree $\leqslant s$, for all large n.

iii) Let $(\tilde{M}_n)$ be another stable $\mathfrak{q}$-filtration of M, and let $\tilde{g}(n) = l(M/\tilde{M}_n)$. By (10.6) the two filtrations have bounded difference, i.e., there exists an integer n_0 such that $M_{n+n_0} \subseteq \tilde{M}_n$, $\tilde{M}_{n+n_0} \subseteq M_n$ for all $n \geqslant 0$; consequently we have $g(n + n_0) \geqslant \tilde{g}(n)$, $\tilde{g}(n + n_0) \geqslant g(n)$. Since g and $\tilde{g}$ are polynomials for all large n, we have $\lim_{n\to\infty} g(n)/\tilde{g}(n) = 1$, and therefore g, $\tilde{g}$ have the same degree and leading coefficient. ∎

The polynomial $g(n)$ corresponding to the filtration $(\mathfrak{q}^n M)$ is denoted by $\chi_{\mathfrak{q}}^M(n)$:

$$\chi_{\mathfrak{q}}^M(n) = l(M/\mathfrak{q}^n M) \qquad \text{(for all large } n\text{)}.$$

If $M = A$, we write $\chi_{\mathfrak{q}}(n)$ for $\chi_{\mathfrak{q}}^A(n)$ and call it the *characteristic polynomial* of the $\mathfrak{m}$-primary ideal $\mathfrak{q}$. In this case (11.4) gives

Corollary 11.5. *For all large n, the length $l(A/\mathfrak{q}^n)$ is a polynomial $\chi_{\mathfrak{q}}(n)$ of degree $\leqslant s$, where s is the least number of generators of $\mathfrak{q}$.* ∎

The polynomials $\chi_{\mathfrak{q}}(n)$ for different choices of the $\mathfrak{m}$-primary ideal $\mathfrak{q}$ all have the same degree, as the next proposition shows:

Proposition 11.6. *If A, $\mathfrak{m}$, $\mathfrak{q}$ are as above*

$$\deg \chi_{\mathfrak{q}}(n) = \deg \chi_{\mathfrak{m}}(n).$$

Proof. We have $\mathfrak{m} \supseteq \mathfrak{q} \supseteq \mathfrak{m}^r$ for some r by (7.16), hence $\mathfrak{m}^n \supseteq \mathfrak{q}^n \supseteq \mathfrak{m}^{rn}$ and therefore

$$\chi_{\mathfrak{m}}(n) \leqslant \chi_{\mathfrak{q}}(n) \leqslant \chi_{\mathfrak{m}}(rn) \text{ for all large } n.$$

Now let $n \to \infty$, remembering that the χ's are polynomials in n. ∎

The common degree of the $\chi_{\mathfrak{q}}(n)$ will be denoted by $d(A)$: in view of (11.2) this means we are putting $d(A) = d(G_{\mathfrak{m}}(A))$ where $d(G_{\mathfrak{m}}(A))$ is the integer defined earlier as the pole at $t = 1$ of the Hilbert function of $G_{\mathfrak{m}}(A)$.

DIMENSION THEORY OF NOETHERIAN LOCAL RINGS

Let A be a Noetherian local ring, $\mathfrak{m}$ its maximal ideal.

Let $\delta(A)$ = least number of generators of an $\mathfrak{m}$-primary ideal of A. Our ambition is to prove that $\delta(A) = d(A) = \dim A$. We shall achieve this by proving $\delta(A) \geqslant d(A) \geqslant \dim A \geqslant \delta(A)$. (11.5) and (11.6) together provide the first link in this chain:

Proposition 11.7. $\delta(A) \geqslant d(A)$. ∎

Next we shall prove the analogue for local rings of (11.3). Note that this proof uses the strong version of the Artin–Rees lemma (not just the topological part).

Proposition 11.8. Let A, $\mathfrak{m}$, $\mathfrak{q}$ be as before. Let M be a finitely-generated A-module, $x \in A$ a non-zero-divisor in M and $M' = M/xM$. Then

$$\deg \chi_{\mathfrak{q}}^{M'} \leqslant \deg \chi_{\mathfrak{q}}^{M} - 1.$$

Proof. Let $N = xM$; then $N \cong M$ as A-modules, by virtue of the assumption on x. Let $N_n = N \cap \mathfrak{q}^n M$. Then we have exact sequences

$$0 \to N/N_n \to M/\mathfrak{q}^n M \to M'/\mathfrak{q}^n M' \to 0.$$

Hence, if $g(n) = l(N/N_n)$, we have

$$g(n) - \chi_{\mathfrak{q}}^{M}(n) + \chi_{\mathfrak{q}}^{M'}(n) = 0$$

for all large n. Now by Artin–Rees (10.9), (N_n) is a stable $\mathfrak{q}$-filtration of N. Since $N \cong M$, (11.4) iii then implies that $g(n)$ and $\chi_{\mathfrak{q}}^{M}(n)$ have the same leading term; hence the result. ■

Corollary 11.9. If A is a Noetherian local ring, x a non-zero-divisor in A, then $d(A/(x)) \leqslant d(A) - 1$.

Proof. Put $M = A$ in (11.8). ■

We can now prove the crucial result:

Proposition 11.10. $d(A) \geqslant \dim A$.

Proof. By induction on $d = d(A)$. If $d = 0$ then $l(A/\mathfrak{m}^n)$ is constant for all large n, hence $\mathfrak{m}^n = \mathfrak{m}^{n+1}$ for some n, hence $\mathfrak{m}^n = 0$ by Nakayama's lemma (2.6). Thus A is an Artin ring and $\dim A = 0$.

Suppose $d > 0$ and let $\mathfrak{p}_0 \subset \mathfrak{p}_1 \subset \cdots \subset \mathfrak{p}_r$ be any chain of prime ideals in A. Let $x \in \mathfrak{p}_1$, $x \notin \mathfrak{p}_0$; let $A' = A/\mathfrak{p}_0$, and let x' be the image of x in A'. Then $x' \neq 0$, and A' is an integral domain, hence by (11.9) we have

$$d(A'/(x')) \leqslant d(A') - 1.$$

Also, if $\mathfrak{m}'$ is the maximal ideal of A', $A'/\mathfrak{m}'^n$ is a homomorphic image of $A/\mathfrak{m}^n$, hence $l(A/\mathfrak{m}^n) \geqslant l(A'/\mathfrak{m}'^n)$ and therefore $d(A) \geqslant d(A')$. Consequently

$$d(A'/(x')) \leqslant d(A) - 1 = d - 1.$$

Hence, by the inductive hypothesis, the length of any chain of prime ideals in $A'/(x')$ is $\leqslant d - 1$. But the images of $\mathfrak{p}_1, \ldots, \mathfrak{p}_r$ in $A'/(x')$ form a chain of length $r - 1$, hence $r - 1 \leqslant d - 1$ and consequently $r \leqslant d$. Hence $\dim A \leqslant d$. ■

Corollary 11.11. If A is a Noetherian local ring, $\dim A$ *is finite.* ■

If A is any ring, $\mathfrak{p}$ a prime ideal in A, then the *height* of $\mathfrak{p}$ is defined to be the supremum of chains of prime ideals $\mathfrak{p}_0 \subset \mathfrak{p}_1 \subset \cdots \subset \mathfrak{p}_r = \mathfrak{p}$ which end at $\mathfrak{p}$: by (3.13), height $\mathfrak{p} = \dim A_{\mathfrak{p}}$. Hence, from (11.11):

Corollary 11.12. In a Noetherian ring every prime ideal has finite height, and therefore the set of prime ideals in a Noetherian ring satisfies the descending chain condition. ∎

Remark. Likewise we may define the *depth* of $\mathfrak{p}$, by considering chains of prime ideals which start at $\mathfrak{p}$: clearly depth $\mathfrak{p} = \dim A/\mathfrak{p}$. But the depth of a prime ideal, even in a Noetherian ring, may be infinite (unless the ring is local). See Exercise 4.

Proposition 11.13. Let A be a Noetherian local ring of dimension d. Then there exists an $\mathfrak{m}$-primary ideal in A generated by d elements $x_1, \ldots, x_d$, and therefore $\dim A \geqslant \delta(A)$.

Proof. Construct $x_1, \ldots, x_d$ inductively in such a way that every prime ideal containing $(x_1, \ldots, x_i)$ has height $\geqslant i$, for each i. Suppose $i > 0$ and $x_1, \ldots, x_{i-1}$ constructed. Let $\mathfrak{p}_j$ $(1 \leqslant j \leqslant s)$ be the minimal prime ideals (if any) of $(x_1, \ldots, x_{i-1})$ which have height *exactly* $i - 1$. Since $i - 1 < d = \dim A =$ height $\mathfrak{m}$, we have $\mathfrak{m} \neq \mathfrak{p}_j$ $(1 \leqslant j \leqslant s)$, hence $\mathfrak{m} \neq \bigcup_{j=1}^{s} \mathfrak{p}_j$ by (1.11). Choose $x_i \in \mathfrak{m}$, $x_i \notin \bigcup \mathfrak{p}_j$, and let $\mathfrak{q}$ be any prime containing $(x_1, \ldots, x_i)$. Then $\mathfrak{q}$ contains some minimal prime ideal $\mathfrak{p}$ of $(x_1, \ldots, x_{i-1})$. If $\mathfrak{p} = \mathfrak{p}_j$ for some j, we have $x_i \in \mathfrak{q}$, $x_i \notin \mathfrak{p}$, hence $\mathfrak{q} \supset \mathfrak{p}$ and therefore height $\mathfrak{q} \geqslant i$; if $\mathfrak{p} \neq \mathfrak{p}_j$ $(1 \leqslant j \leqslant s)$, then height $\mathfrak{p} \geqslant i$, hence height $\mathfrak{q} \geqslant i$. Thus every prime ideal containing $(x_1, \ldots, x_i)$ has height $\geqslant i$.

Consider then $(x_1, \ldots, x_d)$. If $\mathfrak{p}$ is a prime ideal of this ideal, $\mathfrak{p}$ has height $\geqslant d$, hence $\mathfrak{p} = \mathfrak{m}$ (for $\mathfrak{p} \subset \mathfrak{m} \Rightarrow$ height $\mathfrak{p} <$ height $\mathfrak{m} = d$). Hence the ideal $(x_1, \ldots, x_d)$ is $\mathfrak{m}$-primary. ∎

Theorem 11.14. (Dimension theorem.) *For any Noetherian local ring A the following three integers are equal:*

i) *the maximum length of chains of prime ideals in A;*

ii) *the degree of the characteristic polynomial $\chi_{\mathfrak{m}}(n) = l(A/\mathfrak{m}^n)$;*

iii) *the least number of generators of an $\mathfrak{m}$-primary ideal of A.*

Proof. (11.7), (11.10), (11.13). ∎

Example. Let A be the polynomial ring $k[x_1, \ldots, x_n]$ localized at the maximal ideal $\mathfrak{m} = (x_1, \ldots, x_n)$. Then $G_{\mathfrak{m}}(A)$ is a polynomial ring in n indeterminates and so its Poincaré series is $(1 - t)^{-n}$. Hence, using the equivalence of (i) and (ii) in (11.14), we deduce that $\dim A_{\mathfrak{m}} = n$.

Corollary 11.15. $\dim A \leqslant \dim_k (\mathfrak{m}/\mathfrak{m}^2)$.

Proof. If $x_i \in \mathfrak{m}$ $(1 \leqslant i \leqslant s)$ are such that their images in $\mathfrak{m}/\mathfrak{m}^2$ form a basis of this vector space, then the x_i generate $\mathfrak{m}$ by (2.8); hence $\dim_k (\mathfrak{m}/\mathfrak{m}^2) = s \geqslant \dim A$ by (11.13). ∎

Corollary 11.16. Let A be a Noetherian ring, $x_1, \ldots, x_r \in A$. Then every minimal ideal $\mathfrak{p}$ belonging to $(x_1, \ldots, x_r)$ has height $\leqslant r$.

Proof. In $A_\mathfrak{p}$ the ideal $(x_1, \ldots, x_r)$ becomes $\mathfrak{p}^e$-primary, hence $r \geq \dim A_\mathfrak{p} =$ height $\mathfrak{p}$. ∎

Corollary 11.17. (Krull's principal ideal theorem). *Let A be a Noetherian ring and let x be an element of A which is neither a zero-divisor nor a unit. Then every minimal prime ideal $\mathfrak{p}$ of (x) has height* 1.

Proof. By (11.16), height $\mathfrak{p} \leq 1$. If height $\mathfrak{p} = 0$, then $\mathfrak{p}$ is a prime ideal belonging to 0, hence every element of $\mathfrak{p}$ is a zero-divisor by (4.7): contradiction, since $x \in \mathfrak{p}$. ∎

Corollary 11.18. *Let A be a Noetherian local ring, x an element of $\mathfrak{m}$ which is not a zero-divisor. Then* $\dim A/(x) = \dim A - 1$.

Proof. Let $d = \dim A/(x)$. By (11.9) and (11.14) we have $d \leq \dim A - 1$. On the other hand, let x_i $(1 \leq i \leq d)$ be elements of m whose images in $A/(x)$ generate an $\mathfrak{m}/(x)$-primary ideal. Then the ideal $(x, x_1, \ldots, x_d)$ in A is $\mathfrak{m}$-primary, hence $d + 1 \geq \dim A$. ∎

Corollary 11.19. *Let $\hat{A}$ be the $\mathfrak{m}$-adic completion of A. Then* $\dim A = \dim \hat{A}$.

Proof. $A/\mathfrak{m}^n \cong \hat{A}/\hat{\mathfrak{m}}^n$ from (10.15), hence $\chi_\mathfrak{m}(n) = \chi_{\hat{\mathfrak{m}}}(n)$. ∎

If $x_1, \ldots, x_d$ generate an $\mathfrak{m}$-primary ideal, and $d = \dim A$, we call $x_1, \ldots, x_d$ a *system of parameters*. They have a certain independence property described in the following proposition.

Proposition 11.20. *Let $x_1, \ldots, x_d$ be a system of parameters for A and let $\mathfrak{q} = (x_1, \ldots, x_d)$ be the $\mathfrak{m}$-primary ideal generated by them. Let $f(t_1, \ldots, t_d)$ be a homogeneous polynomial of degree s with coefficients in A, and assume that*

$$f(x_1, \ldots, x_d) \in \mathfrak{q}^{s+1}.$$

Then all the coefficients of f lie in $\mathfrak{m}$.

Proof. Consider the epimorphism of graded rings

$$\alpha: (A/\mathfrak{q})[t_1, \ldots, t_d] \to G_\mathfrak{q}(A)$$

given by $t_i \to \bar{x}_i$, where t_i are indeterminates and $\bar{x}_i$ is x_i mod $\mathfrak{q}$. The hypothesis on f implies that $\bar{f}(t_1, \ldots, t_d)$ (the reduction of f mod $\mathfrak{q}$) is in the kernel of α. Assume if possible that some coefficient of f is a unit, then $\bar{f}$ is not a zero-divisor (cf. Chapter 1, Exercise 3). Then we have

$$\begin{aligned} d(G_\mathfrak{q}(A)) &\leq d((A/\mathfrak{q})[t_1, \ldots, t_d]/(\bar{f})) \text{ because } \bar{f} \in \operatorname{Ker}(\alpha) \\ &= d((A/\mathfrak{q})[t_1, \ldots, t_d]) - 1 \text{ by (11.3)} \\ &= d - 1 \text{ by the example following (11.3).} \end{aligned}$$

But $d(G_\mathfrak{q}(A)) = d$ by the main theorem (11.14). This gives the required contradiction. ∎

This proposition takes a simple form if A contains a field k mapping isomorphically onto the residue field $A/\mathfrak{m}$:

Corollary 11.21. If $k \subset A$ is a field mapping isomorphically onto $A/\mathfrak{m}$ and if $x_1, \ldots, x_d$ is a system of parameters, then $x_1, \ldots, x_d$ are algebraically independent over k.

Proof. Assume $f(x_1, \ldots, x_d) = 0$ where f is a polynomial with coefficients in k. If $f \not\equiv 0$ we can write $f = f_s +$ higher terms, where f_s is homogeneous of degree s and $f_s \not\equiv 0$. Apply (11.20) to f_s and we deduce that f_s has all its coefficients in $\mathfrak{m}$. Since f_s has coefficients in k this implies $f_s \equiv 0$, a contradiction. Hence $x_1, \ldots, x_d$ are algebraically independent over k. ∎

REGULAR LOCAL RINGS

In algebraic geometry there is an important distinction between *singular* and *non-singular* points (see Exercise 1). The local rings of non-singular points have as their generalization (to the non-geometric case) what are called *regular* local rings: these are rings satisfying any of the (equivalent) conditions i)–iii) of the next theorem.

Theorem 11.22. Let A be a Noetherian local ring of dimension d, $\mathfrak{m}$ its maximal ideal, $k = A/\mathfrak{m}$. Then the following are equivalent:

i) $G_{\mathfrak{m}}(A) \cong k[t_1, \ldots, t_d]$ *where the t_i are independent indeterminates;*
ii) $\dim_k (\mathfrak{m}/\mathfrak{m}^2) = d$;
iii) $\mathfrak{m}$ *can be generated by d elements.*

Proof. i) $\Rightarrow$ ii) is clear. ii) $\Rightarrow$ iii) by (2.8): see the proof of (11.15). iii) $\Rightarrow$ i): let $\mathfrak{m} = (x_1, \ldots, x_d)$, then by (11.20) the map $\alpha: k[x_1, \ldots, x_d] \to G_{\mathfrak{m}}(A)$ is an isomorphism of graded rings. ∎

A regular local ring is necessarily an *integral domain*: this is a consequence of the following more general result.

Lemma 11.23. Let A be a ring, $\mathfrak{a}$ an ideal of A such that $\bigcap_n \mathfrak{a}^n = 0$. Suppose that $G_{\mathfrak{a}}(A)$ is an integral domain. Then A is an integral domain.

Proof. Let x, y be non-zero elements of A. Then since $\bigcap \mathfrak{a}^n = 0$ there exist integers $r, s \geqslant 0$ such that $x \in \mathfrak{a}^r$, $x \notin \mathfrak{a}^{r+1}$, $y \in \mathfrak{a}^s$, $y \notin \mathfrak{a}^{s+1}$. Let $\bar{x}, \bar{y}$ denote the images of x, y in $G_r(A)$, $G_s(A)$ respectively. Then $\bar{x} \neq 0$, $\bar{y} \neq 0$, hence $\overline{xy} = \bar{x} \cdot \bar{y} \neq 0$, hence $xy \neq 0$. ∎

Hence by (9.2) the regular local rings of dimension 1 are precisely the discrete valuation rings.

It can also be shown that if A is a local ring and $G_{\mathfrak{m}}(A)$ is an integrally closed integral domain, then A is integrally closed. It follows that a regular local ring is integrally closed; but there are integrally closed local domains of dimension > 1 which are not regular.

Proposition 11.24. *Let A be a Noetherian local ring. Then A is regular if and only if $\hat{A}$ is regular.*

Proof. By (10.16), (10.26) and (11.19) we know that $\hat{A}$ is a Noetherian local ring of the same dimension as A and with $\hat{\mathfrak{m}}$ as maximal ideal. Now use (10.22) which asserts that $G_{\mathfrak{m}}(A) \cong G_{\hat{\mathfrak{m}}}(\hat{A})$ and the result follows. ∎

Remarks. 1) It follows from what we have said above that $\hat{A}$ is also an integral domain. Geometrically speaking this means that (locally)

non-singularity ⇒ analytic irreducibility

or that, at a non-singular point, there is only one analytic "branch".

2) If A contains a field k mapping isomorphically onto $A/\mathfrak{m}$ (the geometric case) then (11.22) implies that $\hat{A}$ is a formal power series ring over k in d indeterminates. Thus the completions of local rings of non-singular points on d-dimensional varieties over k are all isomorphic.

Example. Let $A = k[x_1, \ldots, x_n]$ (k any field, x_i independent indeterminates); let $\mathfrak{m} = (x_1, \ldots, x_n)$. Then $A_{\mathfrak{m}}$ (the local ring of affine space k^n at the origin) is a regular local ring: for $G_{\mathfrak{m}}(A)$ is a polynomial ring in n variables.

TRANSCENDENTAL DIMENSION

We shall conclude this brief treatment of dimension theory by showing how the dimension of local rings connects up with the dimension of varieties defined classically in terms of the function field.

Assume for simplicity that k is an algebraically closed field and let V be an irreducible affine variety over k. Thus the coordinate ring $A(V)$ is of the form

$$A(V) = k[x_1, \ldots, x_n]/\mathfrak{p}$$

where $\mathfrak{p}$ is a prime ideal. The field of fractions of the integral domain $A(V)$ is called the field of rational functions on V and is denoted by $k(V)$. It is a finitely-generated extension of k and so has a finite transcendence degree over k—the maximum number of algebraically independent elements. This number is defined to be the *dimension* of V. Now recall that, by the Nullstellensatz, the points of V correspond bijectively with the maximal ideals of $A(V)$. If P is a point with maximal ideal $\mathfrak{m}$ we shall call $\dim A(V)_{\mathfrak{m}}$ the *local dimension* of V at P. We propose to prove

Theorem 11.25. *For any irreducible variety V over k the local dimension of V at any point is equal to* $\dim V$.

Remark. We already know by (11.21) that $\dim V \geqslant \dim A_{\mathfrak{m}}$ for all $\mathfrak{m}$. The problem is to prove the opposite inequality, and for this purpose the main lemma is:

Lemma 11.26. *Let $B \subseteq A$ be integral domains with B integrally closed and A integral over B. Let $\mathfrak{m}$ be a maximal ideal of A, and let $\mathfrak{n} = \mathfrak{m} \cap B$. Then $\mathfrak{n}$ is maximal and* $\dim A_{\mathfrak{m}} = \dim B_{\mathfrak{n}}$.

Proof. This is an easy consequence of the results of Chapter 5. First $\mathfrak{n}$ is maximal by (5.8). Next if

$$\mathfrak{m} \supset \mathfrak{q}_1 \supset \mathfrak{q}_2 \supset \cdots \supset \mathfrak{q}_d \tag{1}$$

is a strict chain of primes in A, its intersection with B is by (5.9) a strict chain of primes

$$\mathfrak{n} \supset \mathfrak{p}_1 \supset \mathfrak{p}_2 \supset \cdots \supset \mathfrak{p}_d. \tag{2}$$

This proves $\dim B_{\mathfrak{n}} \geqslant \dim A_{\mathfrak{m}}$. Conversely given the strict chain (2) we can, by (5.16), lift this to a chain (1) (necessarily strict): thus $\dim A_{\mathfrak{m}} \geqslant \dim B_{\mathfrak{n}}$. ∎

We can now proceed to:

Proof of (11.25). By the Normalization Lemma (Chapter 5, Exercise 16), we can find a polynomial ring $B = k[x_1, \ldots, x_d]$ contained in $A(V)$ such that $d = \dim V$ and $A(V)$ is integral over B. Since B is integrally closed (remark following (5.12)) we can apply (11.26) and this reduces our task to proving (11.25) for the ring B, i.e. for affine space. But any point of affine space can be taken as the origin of coordinates and, as we have already seen, $k[x_1, \ldots, x_d]$ localized at the maximal ideal $(x_1, \ldots, x_d)$ is a local ring of dimension d. ∎

Corollary 11.27. *For every maximal ideal $\mathfrak{m}$ of $A(V)$ we have*

$$\dim A(V) = \dim A(V)_{\mathfrak{m}}.$$

Proof. By definition we have $\dim A(V) = \sup_{\mathfrak{m}} \dim A(V)_{\mathfrak{m}}$. But by (11.25) all $A(V)_{\mathfrak{m}}$ have the same dimension. ∎

EXERCISES

1. Let $f \in k[x_1, \ldots, x_n]$ be an irreducible polynomial over an algebraically closed field k. A point P on the variety $f(x) = 0$ is *non-singular* $\Leftrightarrow$ not all the partial derivatives $\partial f/\partial x_i$ vanish at P. Let $A = k[x_1, \ldots, x_n]/(f)$, and let $\mathfrak{m}$ be the maximal ideal of A corresponding to the point P. Prove that P is non-singular $\Leftrightarrow A_{\mathfrak{m}}$ is a regular local ring.
[By (11.18) we have $\dim A_{\mathfrak{m}} = n - 1$. Now
$$\mathfrak{m}/\mathfrak{m}^2 \cong (x_1, \ldots, x_n)/(x_1, \ldots, x_n)^2 + (f)$$
and has dimension $n - 1$ if and only if $f \notin (x_1, \ldots, x_n)^2$.]

2. In (11.21) assume that A is complete. Prove that the homomorphism $k[[t_1, \ldots, t_d]] \to A$ given by $t_i \mapsto x_i$ $(1 \leqslant i \leqslant d)$ is injective and that A is a finitely-generated module over $k[[t_1, \ldots, t_d]]$. [Use (10.24).]

3. Extend (11.25) to non-algebraically-closed fields. [If $\bar{k}$ is the algebraic closure of k, then $\bar{k}[x_1, \ldots, x_n]$ is integral over $k[x_1, \ldots, x_n]$.]

4. An example of a Noetherian domain of infinite dimension (Nagata). Let k be a field and let $A = k[x_1, x_2, \ldots, x_n, \ldots]$ be a polynomial ring over k in a countably infinite set of indeterminates. Let $m_1, m_2, \ldots$ be an increasing sequence of positive integers such that $m_{i+1} - m_i > m_i - m_{i-1}$ for all $i > 1$. Let $\mathfrak{p}_i = (x_{m_i+1}, \ldots, x_{m_{i+1}})$ and let S be the complement in A of the union of the ideals $\mathfrak{p}_i$.

   Each $\mathfrak{p}_i$ is a prime ideal and therefore the set S is multiplicatively closed. The ring $S^{-1}A$ is Noetherian by Chapter 7, Exercise 9. Each $S^{-1}\mathfrak{p}_i$ has height equal to $m_{i+1} - m_i$, hence $\dim S^{-1}A = \infty$.

5. Reformulate (11.1) in terms of the Grothendieck group $K(A_0)$ (Chapter 7, Exercise 25).

6. Let A be a ring (not necessarily Noetherian). Prove that

$$1 + \dim A \leqslant \dim A[x] \leqslant 1 + 2 \dim A.$$

   [Let $f: A \to A[x]$ be the embedding and consider the fiber of $f^*: \operatorname{Spec}(A[x]) \to \operatorname{Spec}(A)$ over a prime ideal $\mathfrak{p}$ of A. This fiber can be identified with the spectrum of $k \otimes_A A[x] \cong k[x]$, where k is the residue field at $\mathfrak{p}$ (Chapter 3, Exercise 21), and $\dim k[x] = 1$. Now use Exercise 7(ii) of Chapter 4.]

7. Let A be a Noetherian ring. Then

$$\dim A[x] = 1 + \dim A,$$

   and hence, by induction on n,

$$\dim A[x_1, \ldots, x_n] = n + \dim A.$$

   [Let $\mathfrak{p}$ be a prime ideal of height m in A. Then there exist $a_1, \ldots, a_m \in \mathfrak{p}$ such that $\mathfrak{p}$ is a minimal prime ideal belonging to the ideal $\mathfrak{a} = (a_1, \ldots, a_m)$. By Exercise 7 of Chapter 4, $\mathfrak{p}[x]$ is a minimal prime ideal of $\mathfrak{a}[x]$ and therefore height $\mathfrak{p}[x] \leqslant m$. On the other hand, a chain of prime ideals $\mathfrak{p}_0 \subset \mathfrak{p}_1 \subset \cdots \subset \mathfrak{p}_m = \mathfrak{p}$ gives rise to a chain $\mathfrak{p}_0[x] \subset \cdots \subset \mathfrak{p}_m[x] = \mathfrak{p}[x]$, hence height $\mathfrak{p}[x] \geqslant m$. Hence height $\mathfrak{p}[x] =$ height $\mathfrak{p}$. Now use the argument of Exercise 6.]

Index

Printed in the United States
by Baker & Taylor Publisher Services